Adam Phillips

ON WANTING TO CHANGE

Adam Phillips, formerly a principal child psychotherapist at Charing Cross Hospital, London, is a practicing psychoanalyst and a visiting professor in the English department at the University of York. He is the author of numerous works of psychoanalysis and literary criticism, including *On Getting Better*, *Attention Seeking*, *In Writing*, *Unforbidden Pleasures*, and *Missing Out*. He is the general editor of the Penguin Modern Classics Freud translations and a Fellow of the Royal Society of Literature.

ON WANTING TO CHANGE

Adam Phillips

PICADOR

New York

For Seth Phillips

Picador
120 Broadway, New York 10271

Copyright © 2021 by Adam Phillips
All rights reserved
Printed in the United States of America
Originally published in 2021 by Penguin Books, Great Britain
Published in the United States by Picador
First American edition, 2022

Grateful acknowledgment is made for permission to reprint the following
previously published material: Lines from *The World Turned Upside Down*
by Christopher Hill, copyright © Christopher Hill, 1972, 1975, published by
Temple Smith, 1972, Pelican Books, 1975, Peregrine Books, 1984, Penguin
Books, 1991, 1991, 2019, reproduced by permission of Penguin Books, Ltd.
Lines from "Overland to the Islands" by Denise Levertov from *New Selected
Poems* (Bloodaxe Books, 2003). Lines from "Lullaby" by Mary Ruefle from
Indeed I Was Pleased with the World, copyright © Mary Ruefle, 2007,
reprinted with the permission of The Permissions Company, LLC,
on behalf of Carnegie Mellon University Press.

Library of Congress Cataloging-in-Publication Data
Names: Phillips, Adam, 1954– author.
Title: On wanting to change / Adam Phillips.
Description: First American edition. | New York : Picador, 2022. |
    First published in Great Britain by Penguin in 2021.
Identifiers: LCCN 2021037884 | ISBN 9780374172046 (paperback)
Subjects: LCSH: Change (Psychology)
Classification: LCC BF637.C4 P549 2022 | DDC 158.1—dc23
LC record available at https://lccn.loc.gov/2021037884

Our books may be purchased in bulk for promotional, educational, or
business use. Please contact your local bookseller or the Macmillan
Corporate and Premium Sales Department at 1-800-221-7945, extension
5442, or by email at MacmillanSpecialMarkets@macmillan.com.

For book club information, please visit facebook.com/picadorbookclub
or email marketing@picadorusa.com.

picadorusa.com • instagram.com/picador
twitter.com/picadorusa • facebook.com/picadorusa

1   3   5   7   9   10   8   6   4   2

By understanding and cooperating with God's purposes men believed they could escape from the blind forces which seemed to rule their world, from time itself; they could become free.

Christopher Hill, *The World Turned Upside Down*

Let's go — much as that dog goes,
Intently haphazard.

Denise Levertov, *Overland to the Islands*

People pay for what they do, and, still more, for what they have allowed themselves to become.

James Baldwin, *No Name in the Street*

My inability to express myself
Is astounding.

Mary Ruefle, 'Lullaby'

Contents

Preface   ix

Conversion Hysteria   1

Surprise Changes   35

Converting Politics   71

Believe It or Not   105

Coda   139

Acknowledgements   143

Preface

We are changing all the time – growing older and older, whether we want to or not – while often wanting to choose, or even design, the ways in which we change. Now that seasonal change has been displaced by technological change and climate change, and that there seems to be no alternative for the foreseeable future to a capitalist world order, our sense of the changes that are preferable, or even possible, is itself changing. Change may be sought through politics or through therapy, through religion or fitness, through productivity and growth, through relationships or celibacy, or through art and science. But that change is an object of desire goes without saying now; though it is, of course, preferred change that is wanted.

Indeed, one of the things that defines modern societies is the sheer range of invitations to develop ourselves; with the commodification of 'personal growth', there are many options on offer to help people become better at being their best selves (or what they take to be their best selves). If nothing is more revealing of a time, or a culture, or an individual, than their fantasies of change – how they picture and describe the changes that they desire in relation to the changes they take to be beyond their control — we should always be paying attention to how we think about change. And how what we think about change changes over time. Indeed, one of the more puzzling things we have to acknowledge is that how we think about change is changing all the time.

When we think about change now, we have available to us theories of evolution and of natural development (genetics and the life cycle), theories of trauma, histories, political ideologies, religious beliefs, philosophical enquiry, the arts and sciences; and the mixing and matching of all of these in psychoanalysis, and the other so-called psychological therapies. So the story I want to tell in this book — and in its sequel, *On Getting Better* — is about the links that can be made between the traditional, and initially religious, drama of conversion experiences — as a kind of paradigm of profound change for the better — and the usually less dramatic changes envisaged by contemporary psychological therapies, and in particular by psychoanalysis. And about how we got from the language of conversion to a suspicion in the modern era of so-called 'conversion therapies', and, indeed, to larger, contemporary fears about just how intractable or manipulable so-called human nature is.

Conversion was once the profoundest instance of personal and cultural change; and yet it has become, for some people, the most pernicious and disturbing description of the kinds of change we are capable of. Whether it is conversion to religious fundamentalism, communism, profiteering or gender identity, many people now assume that converting people is the worst thing we do (bullying and humiliation are forced conversions). And yet, as we shall see, the idea of conversion is essential to contemporary accounts of personal transformation, just as it was to the beginnings of psychoanalysis, and of William James's pragmatism (James's pragmatism – in which truth is judged by its practical consequences, and not by its sources – and its relation to conversion experience is treated in the final chapter of *On Getting Better*). Both psychoanalysis and American Pragmatism are driven by a desire to help the individual keep things moving. For both Freud and James, the enemy of pleasure and growth was stuckness, addiction, fixity, stasis. They teach us about the temptations of stultification, of the allure of inertia, of the wish to attack our own development; and they suggest, as we shall see, that conversion experiences all too easily become the desire for a change that will finally put a stop to the need for change; change in the direction of what is, to all intents and purposes, a satisfying and reassuring paralysis (converts to religious fundamentalism are not supposed to convert again to something else). They suggest, in significantly different ways, that we are so ambivalent about changing because there is nothing else we can do but change (as though, paradoxically, the fact that we change is the biggest threat to our freedom). And so psychoanalysis and pragmatism try to make wanting to change both appealing and inspiring, as opposed to it being some ineluctable, evolutionary, biological drive, or fate. They promote – American Pragmatism far more than psychoanalysis – the strangely radical, modern political idea that how we want to change can have something to do with how we do change. Change as choice rather than as fate. Change as something we make. And so we need to acknowledge, when we talk about change, the way in which psychoanalysis – and all the other so-called psychological therapies – without American Pragmatism can be merely another coercive, pre-emptive moralism. There are, in other words, two pragmatic questions to be asked of any psychoanalytic or psychological theory, questions the convert always believes she has answered – 'How would my life be better if I believed it?' And 'Would believing it help me get the life I want?' And these questions, of course, lead to further conversations about what my criteria are for a good life, and where I got them from; all underwritten by the idea of the unconscious, of what our knowing and our wanting are really up against. We are always involved, or should be, in the giving of and asking for good reasons to change.

So this book investigates the whole idea of conversion experiences, including accounts of both their history and their allure. It explores both how and why conversion has been such a useful picture of, and analogy for, processes of transformation (but always bearing in mind the historian Hugh Trevor-Roper's remark that 'there is no such thing as a clean break'). In *On Getting Better,* a sequel and complement to this book, I look at the equivalent, for psychoanalysis, and not only for psychoanalysis, of the traditional religious conversion experience, which is the whole notion of cure. And then some of the available modern self-cures are considered – the cure by avoidance ('On Not Having Experiences'), the cure by pleasure ('Unsatisfying Pleasures') and the cure by truth ('The Truth of Psychoanalysis'). Avoidance, pleasure and truth are all deemed to be, in different ways, curative, and integral to the ways in which we want to change; and so they give us the opportunity to wonder what about ourselves may be in need of cure (or in need of pleasure), and why 'cure' seems to be the word for what, in certain circumstances, we think we want. *On Getting Better* ends with a final chapter on William James ('Loose Change') and his way of using the idea of conversion; and how he found a way of breaking its spell, with a view to making us think differently about the changes we want and why we might want them.

Each chapter of both books, then, is intended as at once a free-standing essay, and a linked episode in a story about stories about change (episodes may be more useful than stages – or even development and growth, or continuities in general – as ways of describing the changes we do and don't want). When we talk about change, we are talking about our preferred norms, the standards we want to live by, what we would prefer to think of as normal. So there can be no non-normative therapies, no therapies that don't want to tell us how to live and who to be (and which kinds of change are to be preferred). It is just a question of which of these norms we prefer and why. Wanting to change is as much about our wanting, and how we describe it, as it is about the changes we want. Getting better means working out what we want to get better at.

When we want to think of our lives as progress myths, in which we get better and better at realizing our so-called potential; or conversely as myths of degeneration – as about decay, mourning and loss (ageing as the loss of youth, and so on) – we are also plotting our lives. Giving them a known and knowable shape and purpose; providing ourselves with guidelines, if not blueprints, of what we can be and become. It is not that our lives are determined by our descriptions of them; but our descriptions do have an effect, however enigmatic or indiscernible it might be. And there is no description of a life without an account of the changes that are possible within it. There are, that is to say, the stories we tell about change, and how we actually do change, and they don't always go, or come, together. This book, and its sequel, are about that fact.

Conversion Hysteria

I

What is heartening about people is the appalling stubbornness and the strong roots of their various cultures, rather than the ease with which you can convert them and make them happy and good.

William Empson, review of W. H. Auden's *Another Time*,
in *Life & Letters Today*, August 1940

On 2 October 2012, the *Guardian* reported that 'Britain's biggest professional body for psychotherapists' – the British Association for Counselling and Psychotherapy, which has nearly 30,000 members – had 'instructed members that it is unethical for them to attempt to "convert" gay people to being heterosexual, formalizing a policy change long demanded by rights groups'. Clearly if this had been long demanded by rights groups it had been an issue for some time; there were therapists who believed both that gay people could and should be converted, and that what they should be converted to was heterosexuality; that is to say, there were people who believed that heterosexuality was itself something people could be converted to.

The BACP had written to its members, the *Guardian* reported, to inform them of the new guidelines; the official letter said that the BACP 'opposes any psychological treatment such as "reparative" or "conversion" therapy, which is based upon the assumption that homosexuality is a mental disorder, or based on the premise that the client/patient should change his/her sexuality'. The letter added that it was the World Health Organization policy that, as the *Guardian* put it, 'such therapies can severely harm an individual's mental and physical health'. Once again we must assume that larger powers are invoked – the World Health Organization – because there was a long-standing problem about 'conversion' therapies; that there were a sufficient number of therapists who still believed that homosexuality was a mental disorder, and that people with mental disorders could be converted. We should note too the pairing of 'reparative' and 'conversion' therapies, both intimating, as they do, that something has gone radically wrong, that something needs mending, that wrong paths have been taken; and that somebody knows what the right path is; the ideas of reparation and original sin necessarily conjoined, with conversion as their traditional complement.

Conversion, in its religious context, is usually deemed to be the repair of something; though, as we shall see, it is a word that portends many different kinds of change, all radical but not all reparative. It is a significantly mobile and adaptable concept, a word that can be used (converted) in many contexts – economic, scientific, psychological. Buildings, currencies and energies can be converted. And as we shall see, for Freud, in the very beginnings of psychoanalysis, conversion and sexuality were necessarily linked. So this chapter is about what we are talking about when we talk about conversion, and why we feel as we do about it. Or, to put it the other way round, this chapter is wondering how we can tell, when we change, whether or not we are being converted, and what that might involve as a picture of how we change and how we are changed. What kind of change is inevitable and what kind of change is possible in a life? We are the only animals for whom radical change can be an object of desire. And we are traditionally at our most ambivalent about objects of desire.

So we do need to wonder, here, what so-called sexual orientation is assumed to be, or to be like, if conversion can be the treatment of choice. And, of course – and this will be one theme in this book – how we picture, how we imagine, the conversion process such that a person is transformed from one form of life into another quite different form of life; how a set of founding beliefs can be replaced by another, apparently more persuasive or convincing or compelling – it is difficult to know what the word always is – set of beliefs, in the full acknowledgement that it is always the preferred thing that people are converted to. People are only converted from and to the things that apparently matter most to them – their sexuality here, though once it would have been their religious convictions, these things now blurred and over-lapping, flourishing or not in the same hedgerow. Indeed, the idea of conversion raises fundamental questions about what it is for something to change into something else, and for someone to change into somebody else; involving, as it does, the basic assumption that a person must be something – something recognizable, identifi-able, discernible – in order to be so changed.

Conversion, then, is never less than serious; sup-posedly not real if casual or fleeting, or lightly entered into. We talk of serial monogamists, and serial killers, but we don't talk of serial converters. And yet conversion, which was once, quite recently, one of our most socially sanctioned forms of personal transformation, has itself become one of our most suspect; forcing us to think about just what kinds of personal change we can value and why; what kinds of change we deem to be desirable, and what our criteria might be for forms of personal transformation that we can support and endorse. (That is, what we are willing to let people do to each other.) In short, what kinds of influence we want people to have on each other. From Hitler's Germany and Stalin's Rus-sia to Maoist training camps and radical Islam, we have had in the more modern era apparently horrifying examples of the wish and the will to convert whole nations, to make new men and women. And these examples have left us duly sceptical not merely about the possibility of radical change, but about the desire for it. Are those seeking conversion in some sense, by definition, as it were, lacking, or even ill? What does calling them ill, or deprived, or depraved add to the conversation? And what could they be deprived of that conversion will assuage or appease? What kinds of frustration are conversion experiences a self-cure for? Conversions to radical Islam now, for example, only have a bad press – other, that is, than among radical Islamists – but it is both the religion itself and the conversion process that we have become disturbed by. Would we feel better if these young jihadi men and women had slowly converted during a three-year course at Oxford University in Islamic studies? Why would incremental, evolutionary change be preferred to revelation or revolution?

Clearly a lot depends on how people change and are changed, and what kinds of transformation any given culture promotes; and, by the same token, what kinds of transformation it contests and disparages. What kind of relationship, for example, could a student have to the literature they are studying that would make their teachers or their peers uneasy? Or, in what sense could you be converted to the multiplicity of voices and texts, which both resist and invite interpretation, and that we call literature (you could far more easily be converted to psychoanalysis)? So we might wonder, say, in the literature department of a university, that if we don't want students of literature to be converted to the study of literature, whatever that might entail, what kind of experience do we want them to have? If a student of literature ends up wanting to be an academic, or a writer, in what sense have they been converted and in what sense has something else happened? And how can we describe that something else? What is it not to be converted to something, but to feel it to be in some sense one's vocation? The study of literature, after all, was for some people expected to be a substitute for, or a replacement of, religious belief (as were some of the more secular therapies); and the two things a religious sensibility is most mindful of are the temptations of heresy and the possibilities of conversion. So these are the kinds of questions that a consideration of conversion experiences might provoke.

Liberals, broadly speaking, prefer education to conversion – often intimating that one is the antidote to the other – and therefore they/we prefer conversation to rote learning, multiple perspectives to exclusive explanation, dissent to conformity; and, sometimes, description to explanation. It is, I think, integral to liberal societies to assume that education and conversion are distinct, if not actually at odds with each other. That in liberal cultures people are not educated with a view to their conversion, despite the ineluctable paradox that they are perhaps converts to liberalism. John Stuart Mill presents

8

in *On Liberty* what has become the classic liberal position:

> That mankind are not infallible; that their truths, for the most part, are only half-truths; that unity of opinion, unless resulting from the fullest and freest comparison of opposite opinions, is not desirable, and diversity not an evil, but a good, until mankind are much more capable than at present of recognizing all sides of the truth, are principles applicable to men's modes of action, not less than to their opinions. As it is useful that while mankind are imperfect there should be different opinions, so it is that there should be different experiments of living; that free scope should be given to varieties of character, short of injury to others; and that the worth of different modes of life should be proved practically, when any one thinks fit to try them.

I think it would be true to say that a conversion is not felt to be what Mill calls an experiment of living. And indeed that those who seek to convert others don't think of themselves converting others to half-truths (people convert to things that they believe smack of infallibility). That, in short, conversion tends to be a cure for scepticism; a narrowing of the mind that frees the mind. Frees it, one might say, from the kind of complexity – the relishing of complication, of diversity, of contradiction – that Mill believes makes up the good liberal life (it is noticeable that he uses conversion sparingly in *On Liberty*, and always in a secular context: he refers, for example, to it being 'important to give the freest scope possible to uncustomary things, in order that it may in time appear which of these are fit to be converted into customs'; converting uncustomary things into customs suggesting that, as is often the case, conversion is in the direction of permanence, of new-found continuity). That whatever else a liberal ethos, a liberal education, is taken to be, it is not taken to be a conversion experience, or conducive to a conversion experience; and certainly not a conversion experience to liberal values. It is, that is to say, usually taken for granted that liberalism is, by definition, not something anyone is converted to. And yet, of course, we might think of ourselves, as I say, as having been converted, by liberalism, to being distrustful of conversion experiences. Indeed, in many ways, a liberal arts education aims to arm us against conversion as an object of desire.

Liberals believe that freedom is born of acknowledged eccentricity, complexity and nuance; that, as Mill puts it, 'different persons also require different conditions for their spiritual development; and can no more exist healthily in the same moral, than all the variety of plants can in the same physical, atmosphere and climate'. Mill almost proposes that all consensus is forced consensus; consensus as something we should always be suspicious of (how like-minded can the like-minded ever really be?). He wants to make us suspicious of our apparently natural wish to have things in common, as though agreement is an all too easily disguised collusion, and shared interests are forms of willed compliance. Indeed, the liberalism that he promotes encourages us to ask not what do we have in common, but what do we want to have in common, and why? And what are we using having things in common to do?

Conversion converts us to a new shared world, with new shared purposes. Conversion is always conversion to a group. So what is at stake when we talk about conversion and the kind of hysteria it can both cure and evoke is what people should be doing together, and how people should influence, should affect each other, and what about them should be so influenced; what Mill wants in *On Liberty* is people 'less capable . . . of fitting themselves, without hurtful compression, into any of the small number of moulds which society provides in order to save its members the trouble of forming their own character'. He wants to make us wary of our temptation to fit ourselves into any of the small number of moulds society provides (with the implication that societies only and always provide a small number of moulds); and for the more secular liberal, conversion experiences of virtually any kind would simply be one of the small number of moulds available (Mill, of course, mindful of the pun on moulds, and the implication of the desiccations of conformity). But the converted, of course, do not think of themselves as having been compressed, or fitted into a mould; they would think of themselves as having been released, or enlightened, or liberated. Conversion, that is to say, always involves an essentialism of one form or another. The converted know, in some absolute sense, what is of value, and how it should be valued. They are living in a new truth that renders all other supposed truths false and misleading and corrupting. For the converted, the unconverted are always failing at something and harming something. For those heterosexual therapists who practised conversion therapy, homosexuality was damaging in the fullest sense.

So the British Association for Counselling and Psychotherapy was firmly on the side of Mill's liberalism in the statement by its board of governors. 'BACP believes', they wrote, 'that socially inclusive, non-judgemental attitudes to people who identify across the diverse range of human sexualities will have positive consequences for those individuals, as well as for the wider societies in which they live. There is no scientific, rational or ethical reason to treat people who identify within a range of human sexualities any differently from those who identify solely as heterosexual.' Like Mill, the BACP believes that not only the individual but his whole society is the beneficiary of diverse sexualities, this being itself a judgement despite its promotion of supposedly 'non-judgemental attitudes'. Conversion therapies are opposed to diversity. They are, the *Guardian* journalist Peter Walker writes, 'mainly associated with evangelical Christian groups in the US. It was long presumed that most UK counsellors and psychotherapists recognized that these were widely discredited. But a 2009 survey of 1300 therapists, psychoanalysts and psychiatrists found that more than 200 had attempted to change at least one patient's sexual orientation, and 55 said they were still offering such a therapy.' We need to be able to distinguish pathologizing from bullying.

Perhaps we should not be quite so surprised that the whole notion of conversion, in some ways so very traditional and familiar to us in its (religious) engagement with suffering, has held its ground even in the apparently liberal, secular therapies. Nor indeed that sexual orientation should be treated as akin to religious orientation; as though for many modern people their sexual orientation is their contemporary equivalent of a religious identity.

Philip Hodson, a spokesperson for the BACP, said after one of its members had been struck off for offering 'conversion therapy': 'To me as a therapist it seemed inconceivable that someone who had been trained and made accountable could act in that way. I was shocked rigid that a member was practising conversion therapy, which I thought only happened in wackier parts of America.' We might wonder what Hodson's outrage is seeking to protect here, quite what it is that he needs to distance himself and the BACP from. Presumably 'the wackier parts of America'; and something about how a training that makes people 'accountable' could find itself including conversion therapy within its professional practice. Certainly no therapy today, within secular liberal societies – societies in which there has been what Philip Rieff memorably called 'the triumph of the therapeutic' – that offered conversion experiences would be considered to be anything other than disreputable, so far has conversion fallen from grace as a good story about how people should change each other, or about people changing or, in Mill's terms, forming their own character. 'In general,' Freud wrote in 1920, undoubtedly with some irony, 'to undertake to convert a fully developed homosexual into a heterosexual does not offer much more prospect of success than the reverse.'

We have very different stories now about acceptable forms of change; and indeed about the effects people should be permitted to have on each other. The difference, say, between violation and persuasion. And we have become extremely suspicious about how people can undermine each other – seduce, manipulate, exploit – through words and associated intimidation, as though our language has become potentially the most dangerous thing about us. All the suspicions about psychoanalysis and the other, related talking therapies are all worries about conversion (or brainwashing), with supposedly more rational forms of treatment proffered as alternatives. In this drama, rationality or various fantasies of empiricism are often proposed as our best defence against conversion experiences. As though what is being acknowledged is that something is needed to protect us from whatever it is about ourselves that is prone to conversion. As though conversion was akin to seduction, which it is.

That we are capable of being converted is taken to be the problem. But what exactly is that a problem about? Our receptivity, our longing both for change and for intimacy: our desire, in short, to join other people in something shared, and that makes our lives worth living? Or an essentially megalomaniac drive to be connected to the truth, and so to acquire an inner superiority? Clearly the kind of problem conversion is deemed to be takes us to the heart of something significant: that we do not know what to make of the fact that people can have such a powerful effect on each other; and that we desire these effects as well as fear them. The suggestible, the too easily impressed, the overemotional – the fans and followers and devotees – need help with their susceptibilities.

There is in this, whatever else there is, a terrified misogyny; and a terror of our earlier, more dependent selves. A terror of something about love, and a terror about what the loss of love exposes. There is, in the psychoanalytic story – conversion and psychoanalysis having always been somehow connected – the woman who first, and hopefully often, converted us – the mother who was, in Christopher Bollas's phrase, our first and formative 'transformational object', the woman who, through her care, could radically change our mood; and ourselves as infants and young children desiring and depending on such benign conversion experiences as were possible. 'The mother is experience', Bollas writes in *The Shadow of the Object*, 'as a process of transformation, and this feature of early existence lives on in certain forms of object-seeking in adult life ... The memory of this early object-relation manifests itself in the person's search for an object (a person, place, event, ideology) that promises to transform the self.'

This ongoing search for an object that 'promises to transform the self' is then a reminder of our earliest, most absolute states of dependent need. An object that we depend upon can only be an object that calls up in us the most profound ambivalence. Conversion experiences all too easily, then, have a mixed but not actually a bad echo, both historically and personally. We want to get over them, and we don't. We crave them, and we fear their failure or their unavailability. They link us to our losses, and they remind us of extraordinary boons and benefits. We crave them as opportunities and we fear them as tyrannies. Conversion experiences that, by definition, seem to answer so many questions for the converted can't help, by the same token, but raise so many more questions for the unconverted. The converted, that is to say, are always a provocation to the unconverted, and vice versa.

What then is wrong with conversion? If you feel yourself to be homosexual and don't want to be, why is a conversion experience inappropriate, rather than just the job? One answer would be: sexuality is not the kind of thing that can be converted (any more than friendship is); it is the wrong picture of what sexual desiring is like. Which also allows us to wonder how we do imagine our sexuality if it is not something that we can be converted out of or converted into; if, that is to say, sexuality is unlike sinfulness, or unlike a wrong set of beliefs, or a religious way of life. If we can't convert our sexuality, what can we do with or about it? (Freud, as we shall see, did believe that we could convert our sexual desire into all sorts of other things: but with a view to sustaining our pleasure.)

What do we imagine people are doing to people when they convert them that makes us so fearful of the whole idea? One answer would be: we imagine people willingly being colonized, undermined, overwhelmed, enslaved and having no redress: we imagine people reduced to states of helpless submission, but without realizing that that is what is happening; indeed, being so unconscious of their compliance that they welcome it, they desire it. In this picture conversion is like an apparently benign version of being driven mad, of losing one's mind: the converter and the converted like a sado-masochistic couple, entranced by their ritual: members of a cult that believes it is not a cult, but the truth about life. And what do we think people might be like, what kind of creatures are human animals, if they are prone (unlike all other animals) to being converted? Again, one answer would be that people can't bear certain kinds of frustration, and crave certain kinds of love, or connection to what are taken to be sources of life. That our willingness to be converted is a measure of our abjection, of our need and of our isolation; or of just how unmoored we have become in the cultures we were born into. And, indeed, what do we think language is like, language being the primary medium of conversion, if it can have this kind of effect on people (language also being the medium of psychoanalysis and all the other talking therapies)? And one answer would be that, consciously or unconsciously, we think of language as daemonic. We think of ourselves as doing things with words, while language does things to us.

So the fact that there is such a thing, following on from the language of religion, as 'conversion therapy' at all, however discredited, invites us to consider what might be going on, in this case, in a therapy, if it is not a kind of conversion experience; what kind of conversation can make a difference to people, a difference they desire, that is not a form of conversion? And this, of course, is a version of more ancient misgivings about rhetoric, and about there being something pernicious about the wish to persuade people; or rather to persuade people by disarming them in some way. As though all forms of persuasion involved some kind of disarming; as though all conversation was more or less tentative seduction; and that what we should fear most about ourselves is just how seducible we are (imagine what a world would be like in which our greatest wish was to be seduced). As though there is a magic of words that exploits credulity and insidiously coerces assent. The word 'conversion' itself breaks down into a con version, 'con' meaning 'to know, learn, study carefully' or 'to swindle, trick, to persuade by dishonest means'. We need to think about what honest persuasion might entail. I think psychoanalysis is best described as a form of honest persuasion. Or that, at least, is what it aspires to be.

II

The solidity of undisplaced things, as of selves which have not experienced displacement, may indeed be the greatest of illusions.

Richard Sennett, *The Foreigner*

Appropriately enough, perhaps, the idea of conversion, of course – albeit in a different sense – was integral to the beginnings of psychoanalysis, so-called 'conversion hysteria' being one of the diagnostic categories that Freud (and Breuer) began psychoanalysis with, and began with, in their studies in hysteria; the apparent conversion, as they called it, of affects or ideas into bodily symptoms: conversion as a way of making the unbearable bearable, the unacceptable acceptable enough. Freud writes in an early paper, *The Neuro-Psychoses of Defence* (1894), of the ego

> turning [a] powerful idea into a weak one . . . robbing
> it of the affect – the sum of excitation – with which it is
> loaded. The weak idea will then have virtually no
> demands to make on the work of association. But the
> sum of excitation which has been detached from it must
> be put to another use . . . In hysteria the incompatible
> idea is rendered innocuous by its sum of excitation
> being transformed into something somatic. For this I
> should like to propose the name of conversion.

In this, Freud's first use of the term 'conversion', the
picture is of a 'powerful idea', by which he means the
representation of an urgent but unacceptable desire –
which he will soon refer to as 'forbidden' – that is
stripped of its intensity and rerouted. So my wish – to
take the usual psychoanalytic example – to have intercourse with my mother will be converted into a wish to
enter and explore museums, but I will be troubled by a
pain in my foot as I wander around. Where I was once
possessed by the idea of sexual relations with my
mother, I now say that I really enjoy going to museums,
an innocuous statement of diminished resonance. No
one, including myself, will be upset by this wholly
admirable desire. It is a two-stage process: the exciting,
unacceptable idea is converted into a milder, more

respectable one, and the intensity, the excitement attached to the orginal desire, is displaced into a bodily state, the pain in my foot, which really hurts. 'The neurotic', Freud's colleague Ferenczi wrote in 1912, 'gets rid of the affects that have become disagreeable to him by means of the different forms of displacement (conversion, transference, substitution).'

In these Freudian conversions – 'conversion' being the name Freud chose for these fundamental techniques – the unacceptable idea, the dangerous desire, is displaced. In other words, Freudian conversions involve not a change of heart but a change of means, not a sacrifice of something but a re-presentation of it. What Freud refers to as 'conversion' in his early work is similar to what he will call 'dream-work' in *The Interpretation of Dreams*, which, he writes, 'does not think, calculate, or judge in any way at all; it restricts itself to giving things a new form'. Dream-work is a mechanism, a procedure, an unconscious technique for representing the dream thoughts in acceptable form. Its aim is efficient translation. When desire is converted for Freud, there is a change of form, but not a change of content. 'It is typically Jewish', he wrote to his son Ernst in 1938, 'not to renounce anything and to replace what has been lost.' This reminds us that before these new Freudian conversions – behind them, as it were – was the abiding question of the conversion of the Jews; of what they were willing to sacrifice and renounce in their accommodations to Christian eschatology, and the Christian and Muslim cultures they inhabited.

It is perhaps not incidental that conversion was one of Freud's first and foremost psychoanalytic preoccupations. What he called 'a capacity for conversion' was a capacity to change while remaining the same, a capacity not to renounce anything and replace what has been supposedly lost. 'In neurosis,' Freud's daughter Anna wrote in *The Ego and the Mechanisms of Defence* (1936), 'whenever a particular gratification of instinct is repressed, some substitute is found for it. In hysteria this is done by conversion, i.e. the sexual excitation finds discharge in other bodily zones or processes which have become sexualized.' You don't renounce the sexual desire, you sexualize other areas of your life: instead of being a voyeur, you love reading. Conversion, that is to say – in its psychoanalytic version – is a way of not having to change. It is *the* way the individual sustains the desires that sustain her. So, on the Freudian model, one of the reasons we are so fearful of the conversions to radical Islam – or of any other religious or cultic conversions – is not because they turn people into something they are not, but because they turn people into something that they are. We never change, we convert.

Indeed, in Charles Rycroft's *A Critical Dictionary of Psychoanalysis*, there is an entry for 'conversion', as well as for the more familiar 'conversion hysteria', which reflects the fact that conversion is the heart of the matter for psychoanalysis. 'When used as a technical term,' Rycroft writes, in his definition of 'conversion',

> this refers to the process by which a psychological complex of ideas, wishes, feelings, etc., is replaced by a physical symptom. According to Freud (1893) it is the AFFECT attaching to the 'ideational complex' which is converted into a physical phenomenon, not the 'ideational complex' itself. Although Freud's discovery that physical 'HYSTERICAL' symptoms are PSYCHOGENIC was the seminal observation from which PSYCHOANALYSIS developed, the 'conversion' hypothesis is unsatisfactory since it leaves unexplained what is sometimes called 'the mysterious leap' from the mental to the physical. The mystery vanishes if hysterical symptoms are regarded as gestures.

In this picture it is affects, feelings, that are converted into physical symptoms, which Rycroft suggests, ingeniously, should be 'regarded as gestures' (that is, symbolic movements). But then there is ' "the mysterious leap" from the mental to the physical' – how thoughts and feelings are somatized, become bodily or embodied. Translated, as it were, from the mental to the physical, from thoughts in the mind to bodily states (a fear becoming a phobia, say, or pessimism becoming manic energy). And this itself invites us to consider a larger issue that is integral to the idea of conversion, which is:

23

how suitable, how well suited, is the material being converted to the material it is being converted into? Clearly pounds are well suited to being converted into euros, as are houses being converted into flats; but how does my wish to look at naked bodies get converted into a twitch in my eye? How does my wish to hit my father get turned into a paralysis in my arm?

We tend to think of conversion as involving some kind of shared medium, unbelief converted to belief, or one kind of belief converted into another kind. It was indeed the seminal observation from which psychoanalysis developed – that feelings and desires could be converted into their opposite, into apparently unrelated ideas, into physical symptoms. That the ingenuity of conversion was in its enigmatic transformations; that in conversion we are at our most artful. That, as Rycroft writes, 'ideas, wishes, feelings, etc., [can be] replaced by a physical symptom'. So conversion in this psychoanalytic sense replaces one thing with another; it is a form of substitution. But it means that the thing being replaced has not disappeared (the converted Jew is still a Jew, he has just replaced his Judaism with Christian gestures). Conversion, then, in its psychoanalytic sense, is a cover story (so the converted homosexual would just be a homosexual heterosexual). It is a reconfiguring rather than a radical transformation. Indeed, conversion is in the service of sustaining the very thing that is supposedly being replaced. It is one of the ways the human organism manages the danger of desire, and desire here represents the unacceptable at its most threatening; the individual could survive their desire by apparently transforming it into something else. But it is really a form of smuggling; a way of conserving the desire; not modifying it but disguising it (the patient's symptoms are his sexual life, Freud famously remarks). A Jew is always a Jew; and especially when he seems not to be.

The individual invented by Freud – the psychoanalytic patient that he constructed – was both the converter and the converted; what needed to be converted, as in the Judaeo-Christian religions, was forbidden desire, unacceptable affect, and conversion was in the service of psychical survival. It was the alternative to sacrifice. In this sense psychoanalysis, as Freud conceived it, was the process, through a certain kind of verbal exchange, of deconversion, or reconversion (in the case of Dora, in *Fragment of an Analysis of a Case of Hysteria*, Freud remarks that it is 'difficult to create a fresh conversion', the implication being that this could be the aim); the initial self-cure of converting forbidden desire into somatic symptoms had to be, as far as was possible, reversed; a fresh conversion was needed. The bodily symptom had to be redescribed – another kind of conversion – to at least make conscious, if not free, the desire it compromised.

Psychoanalysis was an attempt to recover from necessary, unconscious conversions; all those conversions of everyday life required by the cultures people found themselves in. In pragmatic terms we could say that conversion was a technique, a process, an art form that was to be used in the service of ever more satisfying gratifications, and not merely suffered as a mechanism of adaptation. Conversion as the more or less successful form our hedonism takes (we are extremely adept, for example, at finding sexual solutions to non-sexual problems). And far from being a rupture, a break with the past, in its psychoanalytic version conversion sustains continuity: in a sense we are more successfully our past selves after a conversion than before. What Ferenczi called in a memorable phrase 'transitory conversions' are the story of our life. To sustain our pleasure, our pleasure in life, we have to be very good at converting and being converted.

III

It is one of the secrets in that change of mental poise which has been fitly named conversion, that to many among us neither heaven nor earth has any revelation till some personality touches theirs with a peculiar influence, subduing them into receptiveness.

George Eliot, *Daniel Deronda*

The phrase 'conversion hysteria' has now a rather different connotation, a new life of its own in our present hysteria about conversion in at least a few of its contemporary forms. So we should also, as I say, bear in mind that in the background of the beginnings of Freud's psychoanalysis – his preoccupation with the conversion hysteria of those women who were unable to fit in, those women who were wanting to sustain their pleasure in adverse circumstances – there was the abiding question of the Jews of Freud's generation in Vienna, the question of whether or not to convert to Christianity with all the losses entailed, conversion being the passport to survival, to assimilation (to an extent), and to professional possibilities that Jews were precluded from. And following on from this, like some terrible intimidating consequence, there would always then be the question of whether the Jew who converts is the least trustworthy Jew of all, the ultimate duplicitous opportunist. (The word itself makes us wonder what the con version of conversion might be.) Freud was, after all, suggesting that in conversion hysteria nothing was eliminated; affect was only transformed: the convert was always haunted, indeed driven, by what she had supposedly converted to and from (in many ways, in Freud's account, conversion was a refinement of desire). Of course, if you can be converted to something, you can pretend to be converted (which was effectively what Freud's hysterics were doing; the hysteric was always accused of acting, which she was and she wasn't, though she was genuinely suffering). How do you know whether the convert is acting, or whether his public performance is consistent with his private life? Symptoms are gestures, and gestures are always gesturing at something, or someone. Desiring, in the Freudian story, requires acting.

It is the irony of all essentialisms – and particularly the essentialisms of character – that they can be aped. Because they are by definition defined, they can be imitated, they can be acted, performed. And conversion always brings us up against the question of essences, because people are only ever converted to what they take to be the essential truth. Indeed, at the heart of conversion is always going to be the question: how can you tell a true conversion from a false one? Which is a version of the ironic question: how can you tell a real essence from a false one? Or the pragmatic question: how can you tell whether a conversion has worked or not? Freud would say that you can never quite tell the difference, because there isn't one. These are the wrong questions because they are, by definition, unanswerable. We are supposed not to be able to answer them; we are supposed to be in doubt. Because in the psychoanalytic story nothing can completely replace (or displace) something else; all substitutions are reminders of what they are substitutes for; all displacements betray their origins. Converts are like transvestites: they disguise something the better to display it. They disguise something to draw our attention to something about it, whether or not they know what that is. Conversion, in other words, for the Jews of Freud's generation in Vienna, could be another word for assimilation.

Psychoanalysis would begin with the all too success-ful but ultimately disabling unconscious conversions of conversion hysteria; and with the questions: what is *there* to be converted, and why is 'conversion' the word? Whatever is there to be converted, in the psychoanalytic story, must be essential and is therefore to be conserved; this means, broadly speaking, instinctual desire, and, essentially, incestuous desire. A redescription is required, not an abolition; not a losing of your life in order to find it, but a translation of your life. So the question is: how do you undo a conversion – in this case the conversion of a desire into a bodily symptom – and what is the cost of both the conversion and its undoing? And what are you left with? Or rather, what, if anything, do you put in its place? Can you keep the conversion process going? From a psychoanalytic point of view it is the conversions that frustrate and get stuck – the conversions that need to believe in themselves – that we suffer from. We suffer inordinately when we allow, in Frank Kermode's terms, a fiction to degenerate into a myth; when we believe too absolutely in what Freud would call our sublimations. 'If a man who believes himself to be king is mad,' writes Lacan in *Écrits*, 'a king who believes himself to be king is no less mad.'

It is perhaps not odd that Freud (and his Jewish col-league Breuer) would be wondering what the preconditions were for a conversion, and what the alternatives were for the unconverted. What can unconverted Jews do (in the world) being akin to the question what can unconverted desires do? How, if at all, can they fit themselves into the cultures they find themselves in? Or rather, which are the conversions that work best? All conversion in this story, we should note, is born of fear. In the sciences Freud studied there were conversions of energy and chemical conversions; but in the lives of his Jewish contemporaries there was the very real question of converting from what was called 'the religion of the fathers'; and in the new science of psychoanalysis there was the understanding and explanation of what we call psychosomatic symptoms, and that Freud began by calling conversion hysteria. Instinctual life, Freud realized, had to be somehow assimilated into a lived life in culture. And this was only a problem for the human animal because of the incest taboo; forbidden desires were in need of conversion. That, in fact, was what development was. Conversion, with its long-standing religious significance, which was all too easily translated into the language of science, proved a very adaptable idea. Conversion could be converted for many different uses, and into many different contexts. Which is why we need to look a little more closely at what Freud was initially using the word to do.

Psychoanalysis would go on being an enquiry into the nature of conversion, its genealogy and its function in the individual's life. To resolve a transference – in so far as such a thing was possible – would be to dispel a conversion; but with a view to, as it were, loosening up the individual's capacity for conversion, making him more susceptible to what Freud called 'fresh conversions' and we might call, following Ferenczi, the 'transitory conversions' of everyday life. Because Freud gradually came to realize that growing up in a family – a family in a society – was somehow akin to a conversion experience, or that this was one way it might usefully be described; that childhood was a protracted conversion and reconversion experience that made us all too vulnerable, all too hungry, all too available and willing to seek out such experiences throughout our lives. It was the gradual conversion to belief (and disbelief) in the parents, for example, he would come to believe, that predisposed us for belief in God.

In the very beginnings of psychoanalysis, however, it was the conversions in so-called conversion hysteria, however disabling, that guaranteed the individual's survival in the family – the necessary but deforming adaptations of modern life – that Freud became interested in. It was the capacity to convert and be converted – to convert forbidden desires into psychosomatic symptoms: to be converted to life in the family – that he described as at once the problem and the solution to the rigours of modern life. In growing up there were the conversions sought and the conversions resisted. Freud needed to understand how conversion worked, and what the work was that it was supposed to do. His various studies in hysteria were an enquiry into conversion and its discontents.

But in reading Freud's uses of 'conversion', the term he claimed to have introduced in *Studies on Hysteria*, we should perhaps take seriously, so to speak, the joke he tells in his 1915 paper *Observations on Love in Transference*; a joke that exposes what could be called a more psychoanalytic take on conversion. Freud has been explaining why it is important, in psychoanalytic treatment, that the analyst makes no concessions to the patient's demand for love. Psychoanalysis should be conducted, as he writes, 'abstemiously' because the point about transference love – a love from the past recreated in the analytic situation – is that it is to be understood, interpreted, not enacted. The patient's love for the analyst is, from the analyst's point of view, transference love. Freud, of course, concedes the difficulties of the distinction, acknowledging the sense in which all love may be transference love, and no less real for being so. But then he tells a joke about a priest (not a rabbi). If the analyst were to gratify the patient, he writes, 'the patient would achieve her aims, but the same would not apply' to the analyst. 'All you would have is a recapitulation of the amusing story about the pastor and the insurance agent. At the behest of his relatives a priest is brought in to convert this seriously ill unbeliever before he dies. The conversation goes on so long that the waiting relatives begin to be hopeful. Finally the door of the sickroom opens. The atheist has not been converted, but the pastor goes away insured.'

It is a strange and telling joke because it likens the analytic situation to a potential conversion experience, and reveals that when conversion is the currency it can go either way. So at this moment in the text, Freud reveals, perhaps unwittingly, a doubt about psychoanalysis, and its project. In what sense are the analyst and his patient like a priest and an insurance agent; and which is which? Conversion, Freud suggests, is more unstable, more unpredictable, more volatile than it looks. Who converts whom in psychoanalytic treatment, and why is conversion what is at stake? The person doing the conversion may not be as sure of himself as he thinks. Indeed, from a psychoanalytic point of view he would have placed himself in this authoritative position as a self-cure for intimidating doubt. Is psychoanalysis, Freud seems to be wondering here, just another version of the oldest genre in the book, the insurance policy (religious or otherwise)?

The idea of conversion, that is to say, explained so much for Freud; it was a way of explaining the individual's inner adaptation to instinctual life, and it allowed him to think of psychoanalytic treatment as a resisted conversion experience. The patient who treats the analyst as, in Lacan's phrase, 'the one who is supposed to know' is wanting a conversion experience, which is interpreted rather than acceded to. But then the Freudian analyst is in the uneasy position of having to convert the patient to the idea of transference, of transference love. But we can say now, in a way that Freud couldn't, that the analyst need be neither a priest nor any other kind of insurance agent. And psychoanalysis can be conceived of then as a conversation in which the wish to convert and be converted is resisted. A conversation in which we can find out what is possible between people when conversion is either no longer possible, or desired.

Surprise Changes

I

Except for the fact that there was no solution, the problem was simple.

Mark Kishlansky, *A Monarchy Transformed*

One of our more abiding descriptions is of our lives as double lives, of our living in, or with — in some kind of relation to — two worlds at once; of ourselves, say, in relation to the gods, or to God (and one's life, then, as the nature and quality of this relationship to divinity, fortunate or doomed, saved or damned); or the two worlds of appearance and reality; or the material and the spiritual. And conversion experiences have tended to be described, in the Judaeo-Christian traditions, as a transition from one to the other; and often despite our modern scepticism of such binary contrasts and oppositions. More secular modern doublings may take the form of the private and the public, or the unconscious and the conscious, or the sane and the mad, or simply the unacceptable and the acceptable. Whether it is good and evil, the sacred and the profane, the well and the ill – what the anthropologist Mary Douglas called 'purity and danger' – the person described is torn; defined by her making and her violation of such boundaries. As though the self, by definition, is that which needs to define itself, and to define itself by insisting on what it is not – a double agent with a lot of work to do.

And so the so-called self is what we have come to call, in William James's phrase, 'a divided self'; and after James and Winnicott, a true and false self, or a self in language, in fantasy, but perhaps, or really, no self at all. A self and its absence co-existing, in its most modern form and formulation. A self always, at least, having to manage conflicting and competing versions of itself; a self always having to get its representations of itself right, even while knowing, in the modern way, that they are only representations, pictures and descriptions of something that may only exist in its pictures and descriptions. A self riddled with conflict, having to straddle the contradictions; or, at its most minimal, do something with or about them.

Conversion is one of the words we use, traditionally, for the treating of contradictions, for the resolving of such conflict. Conversion experiences, at their most minimal, minimise, or at least regulate, what are felt to be the essential and most paralysing contradictions of a divided self. And yet, as we shall see, for modern commentators on the two formative conversion experiences in the so-called Western tradition, Paul and Augustine, these conversions simply expose the conflicts they were meant to resolve and clarify. And this tells us something revealing, so to speak, about our modern scepticism about personal change at its most dramatic and significant. This profound modern ambivalence about conversion experiences – mostly but not always from the non-religious – leads to many questions not only about people's relationship to God, but about their relationship to change, to transformation itself; questions about how it occurs, and what it might be for (what it might be in the service of).

Indeed, one of the questions modern critics and commentators tend to ask is whether there was and is such a thing – both historically and psychologically – as conversion at all. Whether the word itself is a misnomer, and therefore radically misleading, giving us the wrong picture of significant change. And this of course leaves us wondering – both the sceptical and the religious – about what kinds of change may be possible for us; and what kinds of change may be desired. Trauma, we should remember, is our contemporary word for malign conversion experiences; for the experiences that change our lives for the worse, and that threaten our capacity for change.

A more contemporary, secular version of this conflicted self I have been describing is the tension between what we now take to be a biological essentialism and what has traditionally been called the problem of free will. For some people now, who think of themselves as floating free of religious concerns, there is the ineluctable biological unfolding of their lives, and there is a life of decision; a life taking its inevitable biological forms, and a life in some senses chosen, or consented to (as an allegory, it would be Darwin – and perhaps Freud – against the existentialists). Instead of a theocracy there is the determinism of biological process and development, the so-called life cycle – learning to walk and to talk, the self-regulation of bodily need, the irruptions of puberty and middle age, the continual dying; all matched differently in different cultures by initiation rites and institutional arrangements that acculturate and contain the irresistible life stages. And then there are, in certain modern societies, what feel like moments of choice, the wanting and the wanting to be, the lure and allure of cultural ideals, of preferred images; the life of purpose and achievement and predilection, the choosing of company and occupation, the developing of taste. The singularity of individual desire, formed from the available cultural resources. The idiosyncratic consequences of the rendezvous of genetic inheritance and inherited environment; the entangling of transgenerational history with available contemporary circumstances. There is the being fashioned – by what we have been taught to call nature and culture – and the ongoing question of self-fashioning; the turning of oneself into who one might want to be. What one has to live with, and how one wants to live. What we do with, or about, or for, the presenting determinisms – the gods, God, nature, instincts, history, economics, genetics, race, language, sexuality, childhood. What we think we know about as the organizing forces of our lives, and what we don't. And whether, or to what extent, we are delegating our agency to these determining powers to absolve ourselves of responsibility. So for the purposes of this chapter, I want to think of a nexus of overlapping preoccupations – determinism, choice, and in the middle, as it were, conversion experiences, which are, whatever else they are, a way of going on talking about the problem of free will; of whether, and in what ways, we can choose our transformations; and how we know they are good; of the constraints on translation – the translation of ourselves – and of where we get our criteria for the good translation.

Instead of the question being, starkly, 'Are we changed or do we change ourselves?', conversion experiences suggest that both may be possible, so the question is wrongly phrased. If change is happening every moment of our lives – at a psychobiological level – and at the same time we are able both to want change and to have the fantasy that nothing is changing; if we are creatures that by nature and therefore by definition change, and creatures who want to choose and design their changing, what then is change experienced as being, or being like? If change is both an object of desire and a fundamental fate, where does this put us? What, if anything, have we got to do with it? Conversion experiences can be both the desired change in a person's life – either a change they were conscious of wanting and hoping for, or a desired change only recognized in retrospect; or it can be an entirely unanticipated transformation that is only of supreme value after the event ('trauma', as I say, being the word we use for malign, unchosen conversion experiences). Change as an object of desire is a question of knowledge, of in some sense knowing what we want to be, or to become, or knowing that we don't know what we want but that we want something; or change as ineluctable biological fate – living through the rewards and ruins of time – and so something we may be able to know about, and to some extent intervene in through the wonders of science, but fundamentally a question of acknowledgement (of the unchosen as a clue to what then may be chosen: if we can't choose not to die, what then, what in the light of that fact, can we choose about how to live?). The rules make improvisation possible, but there can be no rules for improvisation (the improvised being the possible made possible). The rules are, by definition, known beforehand; the improvisation, by definition, cannot be known beforehand. What, if anything, do you need to know to get the life you want, the life you want being the life in which you change according to your wishes?

If the modern question is, as Michel Serres suggests, 'What is it you don't want to know about yourself?', the pragmatic question that follows on from this is: what can you do with whatever it is you do and don't want to know about yourself? One of the things we do and don't want to know about ourselves is how we are unavoidably changing, and how we might want to change. There is, in our more modern, secular languages, the double life of biological destiny and self-invention, and the double life of what you think you know, and want to know, about yourself, and what you don't want to know about yourself (what is conscious and what is, as yet, unconscious). What is being converted (and apparently cured) in conversion experiences are contradictions such as these. It is not, though, that the problem is solved; it is that to all intents and purposes it disappears. It ceases to be the issue that it was. As often happens in a psychoanalysis, people aren't cured, they just lose interest in their symptoms. Their preoccupations fade, and evolve. The converted do and don't have different preoccupations.

The phrase 'double life' has, of course, for us – or for those of us who live in a more secular world – the connotation of infidelity, and so of the very real difficulties of wholeheartedness; of being of a piece in our desire. There are people today called, broadly speaking, liberals – or fans of John Stuart Mill's *On Liberty* – who do not believe that purity of heart is to will or love one thing; or indeed that there is such a thing as purity of heart (or such a thing as one thing). These people believe in mixed motives and competing desires, in ambivalence and paradoxical accommodations; they believe that the best is the enemy of the good – that good enough mothering is more than good enough – and that the worst thing we do is intimidate others (intimidation being an attempt at forced conversion); that narrowing our minds – diminishing our and other people's complexity – makes us unduly cruel. And that one way we have narrowed our minds is to describe our lives as only, or merely, double. Doubling doubles the problem in a way that multiplication doesn't.

Doubling, we can say now in the language of psychoanalysis, is a defence against proliferation. And so I want to suggest that wherever and whenever there is modern, as opposed to more traditional, talk of conversion, there is – among many other things that are historically specific – the threat of proliferation. People are prone to conversion experiences when they can no longer bear the complexity of their own minds. We don't want to kill the person we hate most, the psychoanalyst Ernest Jones once remarked; we want to kill the person who precipitates us into the most unbearable conflict.

So I want to suggest that when we talk about conversion, we need to keep two modern questions in mind. What was the unbearable conflict for which conversion was a self-cure? And what exactly was the complexity of mind – the excess of thoughtful feeling – that needed to be simplified by being apparently clarified. Kill them or convert them. We want to convert not the people we hate most, but the people who precipitate in us the most unbearable conflict. We want to be converted by those people who can apparently resolve our most unbearable conflicts. Conversion is a symbolic murder.

We need, then, in considering all this, to look again at two of the most formative conversions in the so-called Western tradition; two examples – Paul and Augustine – that are of course linked, and taken to be at once exemplary and defining. They are among our most dramatic accounts of radical transformation; and by the same token perhaps – even though this is, in part, quite at odds with their intention – they reveal just how enigmatic and puzzling the whole notion of benign catastrophic change has become for us; a consummation devoutly to be wished and feared; and apparently an abiding preoccupation of this Western tradition, so convinced, as it is, of the value of its own preferred forms of transformation as to be determinedly committed to the conversion of others. Converting others being the only antidote to one's own doubt.

Our favourite stories are all about individuals and groups wanting and not wanting to change. But we can read conversion scenes and not be converted by them. It is, though, our transformation scenes that haunt us; but we may not have been curious enough about how these two things go together, this drive to change and be changed, running alongside the changes that are always already happening. The transience that we can only transiently acknowledge and the mourning we are so keen to do have so far been our best conceits and containers for the perennial question of what to do about change. It is not odd, of course, that we have been so inventive about the inflicting and the enduring of suffering; that we have been so imaginative in our turning trauma into triumph. And yet changing in ways we prefer – in Freud's now much-criticized formulation, 'where id was, there ego should be', the emphasis on the should – seems to be the unavoidable human project. What is it for a person to be changed, and to change themselves, for the better, and how does it happen? For every other animal, change takes care of itself. No other animal goes to these particular lengths to avoid suffering and pursue cultural ideals. Or, to put it the other way round, it has been difficult for us to find pleasures that are sufficiently sustaining; pleasures that convince us, so to speak, that our lives are worth the suffering.

It may be useful, then, if a little reductive and distracting, in the modern way, to work out what the pleasures were that Paul and Augustine were seeking, and what pleasures they exchanged for them, though we must be mindful that by definition they could not and would not have put it like this. They could not conceive of themselves, except sinfully, as pleasure-seekers. Perhaps it would be more accurate to say that they were, in secular terms, replacing their objects of desire; which would itself be one definition of a conversion. For both of them, after all, were converted – as the converted always are – to something they believed they could absolutely (and finally) depend upon, and that therefore made their lives unequivocally worth living. To something, that is to say, that apparently cured an unbearable conflict, and clarified, perhaps by simplifying, an excessive, unbearable state of mind.

II

. . . grotesquerie and strangeness may serve as at least a partly converting invitation.

Rowan Williams, *On Augustine*

Any account of Paul's founding and famous conversion on the road to Damascus has to begin with a paradox which is itself of interest. As Rowan Williams writes, for example, in *Meeting God in Paul*, 'The language of conversion from one religion to another was simply not available [in Paul's time] in the sense we give it today'; and yet later in the book he gives an account of what he calls 'the character of Paul's conversion as he and others tell us about it'. Despite the discernible contradiction, the idea of the character of a conversion is worth holding on to. Scriptural scholar Paula Fredriksen elaborates, in a useful way, on the supposed contradiction of the fact that, as she puts it, 'what we call "conversion" was so anomalous in antiquity that ancients in Paul's period had no word for it'; and yet, as she says, Paul 'emerges as history's most famous convert, a sort of honorary gentile Christian avant la lettre and, at the same time, and thereby, an ex- or anti-Jew. Indeed some scholars still argue, Paul stands as history's first Christian theologian, urging a new faith that supersedes or subsumes the narrow Judaism of his former allegiances.'

Paul as history's most famous convert in a culture in which there can be no such thing as conversion; Paul who we have needed to see as having had the paradigmatic Christian conversion, at a time when such a thing, such a description, was impossible, literally unintelligible. It was as absurd as the idea of converting one's personal and transgenerational history, or one's genetic inheritance. 'The modern term', Fredriksen writes, 'for such a transition, "conversion", fits poorly in Paul's period, when one's kinship group, the genos or ethnos, anchored and articulated piety. Given the essentialism of ethnicity in antiquity . . . how did a pagan become a Jew? For some Jews the answer was: impossible. This position privileged antiquity's normative connection of family and cult, and its realist construction of genealogy and "blood".'

Where we talk about conversion in relation to Paul, Fredriksen suggests, we should talk about transition; the essentialism of ethnicity then setting the limits to change (we would now argue about what, if anything, constitutes ethnicity). In this account – which seems to be of a piece with many of our contemporary accounts – Paul was not transformed into somebody else on the road to Damascus; he did not become somebody he was not, but he suffered a dramatic transition (how can somebody become something they are not?). And from this point of view – informed as it must be by Darwinian biology and modern historiography – we are urged to see the continuities and not the rupture in his conversion experience. Where once there seemed to be revolution, now there is punctuation, evolution rather than apocalypse. And so contemporary accounts, both of Paul's experience on the road to Damascus and of his theology, struggle with this fundamental question: was Paul continuing and elaborating the Jewish tradition in which he was brought up and which he practised – and so 'reclaiming more of the Jewish roots of [Christianity]', in Michael Thompson's words – or was he radically revising, or even repudiating, it? This is a version of the question: what kind of change was deemed possible in a person's life then (and now), and how are we to judge its value? No one, after all, says a terrible thing happened to Paul on the road to Damascus, though many people took against his teachings, in the most virulent way. It is a question, as Thompson says, of the 'amount of continuity' between Paul's past and his revelation of a new life. Through his conversion, he went from being a defender of the purity of Judaism to promoting the absolute inclusiveness of the gospel.

Something happened to Paul on the road to Damascus, for which there must have been preconditions, if not discernible causes. And many people, including Paul himself, didn't quite know what to make of the experience. But they did know that it was an experience, and that a great deal was made of it, and indeed came of it. A so-called conversion experience is, then, defined by its effect, by what it prompts and inspires both in the converted and in the people with whom he is in contact. Paul's conversion experience, or transition, was defined by its after-effects; one of which was that it became some kind of scene-of-instruction for Western Christians. So we need to look, albeit briefly, at the experience itself, and some of its consequences. In brief, as most people know, Paul, a virtual contemporary of Jesus, whom he never met, had been a pious, educated Jew – a tent-maker by trade – who had been an avid persecutor of Jesus's new sect. He did not, that is to say, believe that Jesus was the Messiah of the Jews.

And Saul, yet breathing out threatenings and slaughter against the disciples of the Lord, went unto the high priest, and desired of him letters to Damascus to the synagogues, that if he found any of this way, whether they were men or women, he might bring them bound unto Jerusalem. And as he journeyed, he came near

Damascus, and suddenly there shone round about him a light from heaven. And he fell to the earth and heard a voice saying unto him, 'Saul, Saul, why persecutest thou me?' And he said, 'Who art thou Lord?' And the Lord said, 'I am Jesus who thou persecutest: it is hard for thee to kick against the pricks.' And he trembling and astonished said, 'Lord, what wilt thou have me do?' And the Lord said unto him, 'Arise and go into the city, and it shall be told thee what thou must do.' And the men who journeyed with him stood speechless, hearing a voice, but seeing no man. And Saul arose from the earth, and when his eyes were opened, he saw no man; but they led him by the hand.

Saul/Paul begins by 'breathing out threatenings and slaughter' against the followers of Christ; then he is struck down by a light from heaven and a voice; he is asked a question, and then he asks the voice to make demands on him: 'Lord, what wilt thou have me do?' He is blinded – according to another account – he is gradually told what to do, and he recovers his sight: 'there fell from his eyes as it had been scales, and he received sight forthwith, and arose and was baptized'. The vision he recovers, so to speak, is different from the vision he lost; the text does not say he recovered his sight but that he received it. And this, of course, is the all too familiar description of conversion: the loss of an old way of seeing, and a new vision of things. The same eyes see things differently and are at once the old eyes and the new eyes.

But I want to emphasize two things in this description. First, the violence of the experience, which comes from outside; Paul has in no sense sought out or tried to engineer this experience, which initially undoes him; it is, indeed, the experience he is wanting to abolish, to stop people having, the experience of recognizing Jesus as the Messiah of the Jews. And second, the experience involves Paul's requesting a demand, demanding a demand, as it were, 'Lord, what wilt thou have me do?' It is as though he is wanting something new to be asked of him that he cannot ask of himself; or, indeed, be asked by other people. He can only be asked by Christ. Paul gets the opposite of what he thought he wanted. So, in the plain words of theologian John Barclay, 'Paul was an exemplary Jew',

> full of zeal, a first-class student advanced in Judaism beyond his contemporaries. He was also in that zeal, a key figure in persecuting the church, fully justified in so doing by 'the traditions of my ancestors' (Galatians, 1:14). But then he was encountered by a revelation of Christ, since he was set apart from before his birth and called by grace (15/16). That grace was no reward for his excellence in Judaism. In fact, that excellence, he now realized, had led him entirely in the wrong direction ... the grace of God was given to him,

without regard either to his merits or his demerits. God took no account of who he was or what he had done. Neither his ethnic identity, nor his moral successes or failures had any bearing on the definitive gift of God . . . God is making clear that the gift is not given to fitting or worthy recipients as judged by any previous criteria; God is subverting the very criteria by which one might judge worth.

The demand that Paul has demanded is to live a life without knowing what its demands are; to live, in other words, a life of faith and grace and not of works. The one thing you need – your salvation through grace – is the one thing you cannot will or achieve: 'God is subverting the very criteria by which one might judge worth.' Anything created by people – even the very criteria by which one might judge worth – is by definition not to be trusted. Paul, one can say in Lacanian language, has replaced one so-called Big Other with another Big Other. The God of his Jewish tradition demanded of him that he observe the Jewish law; Jesus, God's supposed son, demands that he put his faith in a God whose demands he cannot know. What is revealed to him is that he cannot know what is wanted of him (and therefore quite what he himself wants); and that what he most needs – God's grace – he can do nothing to secure (hope is always hope for the wrong thing). As a Jew you can – at least to some extent – know what is wanted of you; in

Paul's Christianity this is the one thing you can never know. What is demanded of him is a kind of competence without comprehension; the acknowledgement that he has faith in a God he can never know, or know about. This is a new kind of faith that has nothing to do with knowledge. So what is revealed to Paul on the road to Damascus is that he has no agency in his own life, his life is in no sense his own. His conversion is his lived acknowledgement of his absolute dependence, and of this absolute dependence being on God through Jesus (and this is the thing that Augustine will pick up on). The idea of absolute dependence on something other than oneself – or on something other than the human – was, as it were, the traditional revelation. Absolute dependence on one particular God, deemed to be the only – omniscient and omnipotent – God was the news announced in the Old Testament, and brought to fruition in the New Testament. But the question has always been: was Paul's conversion a break or a link, a rupture or a recovery? Is it what Michael Thompson calls a reclaiming of the Jewish roots of Christianity, or a radically new religion?

After Paul, the traditionalists will say: you don't give up on the law now, you don't treat it as something it isn't, the way to God's heart. The revolutionaries will say that the law is the enemy of grace; it is sinful pride, a refuge from faith; it is a kind of arrogance to do your best, as though you could second-guess God and know what he wants (it would be the ultimate omniscience to know God's desire). Paul's conversion, in other words, as his legacy makes clear, elaborated the problems it was trying to solve (it didn't stop commentary, it prompted it). Paul may have seen what he took to be the truth on the road to Damascus, but his texts, and the voluminous commentary they have generated, revealed the ongoing ambiguity and indeterminacy of this truth. Was it a revision, a radical redescription of Judaism, or was it a new religion? Was it a repudiation of a tradition, or a renewal of it? What was being added and what was being given up?

And this brings us to one of the hearts of the matter of conversion, and so of our stories about significant change; and so of our haunting by the story of Paul's conversion. Paul didn't know what he wanted, or that he was wanting anything other than what he already wanted – to observe the Jewish law, and so condemn the Christians. What he discovered, through what came to be known as his conversion, was not that there was no law but that the law was not what he thought it was; what he discovered was not what God's demands were but that he had misunderstood the nature of God's demands, and therefore of God. The old objects, that is to say, had not been abolished, they had been redescribed; much of the old vocabulary was still there – God, faith, law, mercy, charity, grace, and so on – but grace was privileged over law, faith was preferred to observance, and there was more love and mercy in the picture. In other words – it is an obvious point, but one of significant consequence – in conversion experiences there has to be something there to be converted; the new can only be made out of the old. And it may be that one of the ways the new makes a name for itself is through the rhetoric of hyperbole; great claims are made, but over time the past begins to show through. Paul took Abraham to be his precursor (in Galatians 3 and Romans 4, he is clear that justification by faith was true for Abraham, and so not a new idea; he himself asserts in Romans 9 that Israel was originally established through divine mercy and grace). There was always, as he acknowledged, what John Barclay calls, 'the Grace-Shaped story of Israel'. Whatever else they do, conversion experiences always also smuggle in the old in the guise of the new (how do you keep the past alive? By making it new). So we may need to think of conversion also as conservation; as the conservation that has to be denied.

It is worth wondering, then, why Paul's conversion experience has stayed and stuck as Western history's most famous conversion, and as the new beginning of what became a world religion. And why, by the same token, conversion experiences were taken to be of such significance, for some people the only real emblem of significant change. At its most minimal it shows us that we can be changed, that something is committed to our changing (whether it is God or something in ourselves), and that there is something other than ourselves that can change us (Paul's conversion was, as many commentators suggest, a calling; indeed, John Barclay in his book *Paul* suggests that 'calling' in Paul 'is the word he uses for what we term "conversion"'). It also shows us that we do not – perhaps cannot – know the transformations that we seek (beforehand); and that, of course, the changes that do occur themselves have unpredictable consequences. That under the guise of settling something, significant change unsettles everything (nothing breeds uncertainty like certainty does). And perhaps above all, so to speak, that traumatic change can be for the good (something that lived experience could lead one to doubt). Paul's conversion experience was, in the fullest, most cosmic sense, hopeful. Dramatic change and hope were made compatible. The world itself was to be saved. We could be led to believe that the greatest change could lead to the greatest hope – that death itself could be our greatest hope. This, one could say, is the opposite of James Joyce's remark when he was asked what he thought of the after-life: 'I don't think much of this one.'

Conversion experiences, in the Pauline tradition, cannot be contrived, engineered, calculated or taught (they are not born of persuasion); this, one might say, is their subversive power, the subversive power Paul attributed to grace itself, and that we can attribute also, to some extent, to natural processes of growth and decay, and indeed to accident. For Paul, conversion experiences are in God's hands and not in what he called man's (anything made by people, in this story, is corrupted by sin; so politics or scientific method could only ever be fundamentally flawed). But Paul's conversion experience did, as it were, set precedents, provide pictures, or models, or examples of significant change; and of why we might be interested in change beyond the 'natural' changes that we cannot evade or avoid (something in us needs to be redeemed). And this raises questions of where such change might come from, why it might occur, and what its consequences might be. Of course, in the Judaeo-Christian cosmology, there is only one source, by definition, of anything significant (and so of significant change), and that is God. And as we will see later, there are overwhelming consequences of this, for secular stories about how we might change or be changed. One of the more obvious after-effects of God-driven conversion experiences has been the search for other, or alternative, sources of and promptings for change, and for secular criteria for what might be the desirable changes in a person's life But we go on getting the measure of Paul's paradigmatic conversion experience through the conversion experiences that followed on from it, of which Augustine's conversion is our second exemplary instance; taking its cue, as it were, from Paul.

III

There is nothing in literature but change and change is
mockery.
  William Carlos Williams, *Kora in Hell: Improvisations*

In 386, at the age of thirty-two, Augustine, just before
his conversion – which we might call the second-most-
famous conversion in Western history – started reading
Paul's letters. And as we shall see, a further text of
Paul's played a central part in Augustine's actual conversion experience. Paul as precedent or precursor is
important partly because of what we have come to call
acculturation – i.e. how would Augustine have known
it was a conversion experience he was having if he
hadn't, among other things, read Paul? But also because
one of the things that always comes in the wake of conversion experiences, as I say, is a scepticism about them
(which involves some kind of reconstruction of the
reported experience). Just as Paul's conversion, as we
saw, over time made people ask: had Paul developed
Judaism or had he started a new religion? – was his
work what we would call development or revolution,
or development as revolution: something utterly old or
utterly new? – so Augustine's conversion, as seen
through voluminous commentary, has made people

wonder: what, if anything, changed, and how so? Was it really as sudden as it sounded, or was it, in Hopkins's words, 'a lingering-out sweet skill'?

In the double act of conversion, there are always the proclamations and declarations of the converted ('it was as if a light of relief from all anxiety flooded into my heart', Augustine writes in Book VIII of the *Confessions*; 'all the shadows of doubt were dispelled'); and then there is the complementary scepticism of the critics ('The problem with a Christian conversion', Miles Hollingworth writes in his book on Augustine, 'is that you have to wake up from it the following morning, much unchanged'). Yes, dramatic transformation occurs, but it's not quite as dramatic as it looks. It is as though what haunts all religions – and then all secular therapies – is that the wish for change may be in excess of the capacity for change. 'The supreme challenge for any biography of Augustine, for scholarship and the reader,' writes Hollingworth, 'is to make the event of his Christian conversion in the summer of 386 just as much the key to his life story as he was determined to make that life story the key, in turn, to so much of the Western intellectual tradition after him.' A modern psychological, rather than theological, question would be: why is there a wish both for a key to one's life story and a wish that the key would depend upon a dramatic transformation? How do we imagine a life if it requires conversion experiences, or something akin to them?

Whereas with Paul there has been contention, as we saw, about whether 'conversion' was a plausible word in antiquity for his experience, by the time we get to Augustine, no one seems to question whether it is the right word for what happened to him in the garden in Milan. Peter Brown, in his magnificent biography of Augustine, uses the word 'conversion' in inverted commas for Augustine's gradual commitment to philosophy; but he remarks, without inverted commas, that 'in the *Confessions* we have . . . the authentic words of the convert', and he quotes Augustine from Book IX:

> The days were not long enough as I meditated and found wonderful delight in meditating upon the depth of Your design for the salvation of the human race, I wept at the beauty of Your hymns and canticles, and was powerfully moved at the sweet sound of Your Church's singing. These sounds flowed into my ears, and the truth streamed into my heart; so that my feeling of devotion overflowed, and the tears ran from my eyes, and I was happy in them.

Why the 'authentic words of the convert'? Because Augustine has discovered an absolute dependence; he knows now that everything of value about himself and the world comes from God. 'A sense of purpose and continuity', Brown writes, 'is the most striking feature of Augustine's conversion'; his life is now of a piece because of his unassailable acknowledgement of God as the source and the judge. And yet Brown also qualifies this in a way that is now familiar, at least to our modern sense of things. 'The tastes of Augustine's age', he writes, noting the popularity of the genre,

> demanded a dramatic story of conversion, that might have led him to end the *Confessions* at Book IX. Augustine instead added four more, long, books. For Augustine, conversion was no longer enough. No such dramatic experience should delude his readers into believing that they could so easily cast off their past identity. The 'harbour' of the convert was still troubled by storms; Lazarus, the vivid image of a man once dead under 'the mass of habit', had been awoken by the voice of Christ; but he would still have to 'come forth', to 'lay bare his inmost self in confession', if he was to be loosed. 'When you hear a man confessing [Augustine writes], you know that he is not yet free.'

Attributing, I think correctly, these misgivings about the finality of conversion to Augustine himself, Brown (and perhaps Augustine himself) reminds us of the essential question: what happens to the past in a conversion experience? Which is a version of the question: how long is a conversion experience? How long does it take, when does it start and when is it finished? ('No such dramatic experience', Brown writes, 'should delude his readers into believing that they could so easily cast off their past identity.') It is indeed improbable that a past identity could be abolished; and so, if it is not abolished, in what form does it live on? There may be continuities, uncanny or otherwise, between a past identity and the identity after dramatic transformation; more or less subversive or confounding, more or less reassuring and consolidating. Conversion, Brown's Augustine asserts, is an ongoing work; the conversion experience itself initiates something; or rather in Augustine's case returns him to something, recovers something. 'Unlike Paul's,' Robin Lane Fox writes, 'his conversion had been a long slow process, requiring a final leap.'

In one of his first writings after his conversion, Augustine refers to the Christian religion as something that had been 'inserted into us as boys, and had become entwined in our very marrow, our innermost part'. So Lane Fox, in his recent book – significantly entitled *Augustine: Conversions and Confessions*, to suggest that Augustine's was a cumulative conversion – rather than contesting the idea of conversion for Augustine, justifies it etymologically. 'The Latin word *conversio* had a more general scope,' he writes. 'It could mean a returning to God by someone who had turned away (*aversio*). Literally it could refer to the turning round of a congregation in church, as we learn from one of

Augustine's sermons . . . Less familiarly for our English usage, *conversio* could refer to God's own turning to his own universe, or to individuals in it. This sense, a turning towards us by God, is the sense of the word *conversio* when it first occurs in the *Confessions*.'

So we have here, for Augustine, conversion as a return, a turning towards, 'a reverting' after a turning away (his notorious debauchery, rhetorical ambitions, and Manichaeism); but we also have his conversion described by Lane Fox as a 'culminating conversion' – that is to say, the final conversion in a series of conversions that defined his life. In addition, we have, in an echo from Paul, a turning towards the agent of change, who is God. There is, in Augustine's case, unlike Paul's, the desire for conversion (a desire, perhaps, partly evoked by Paul); and like Paul a fundamental helplessness in the willing of it. The change of conversion is utterly dependent on God, on something other than human agency or resources. And this story of change could only be secularized, in modern therapies, through a kind of magical reversal: change as dependent only on human agency and resources. It being, of course, another element of secularization not to replace one omnipotence with another. And once you take omnipotence and omniscience out of the picture – the God that Paul and Augustine relied upon – the changes possible in a life would seem to be up for grabs (it is no longer clear what good change is). It would seem to be not incidental to this larger story – and something I will come back to – that some of the modern psychoanalytic therapies are organized around an understanding of the omnipotent and omniscient part of the self as the saboteurs of change and development.

For Paul and Augustine – as exemplars in the Judaeo-Christian tradition – there is only one change that is of value, and from which all valuable changes come. These Christian conversions set the pace, provide the model, for desirable and significant change in a life. They are, as it were, the presiding geniuses that haunt and inform our apparently secular transformation scenes; our beliefs and assumptions about how to change and be changed. In so far as they are before-and-after stories, they involve quite elaborate narratives of preconditions and consequences; but preconditions and consequences focused on an emblematic event – the conversion experience – that we might also think of as the precursor of modernist epiphanies; those moments that clarify and transfigure what T. S. Eliot – himself a famous convert – called in an essay on Walter Pater, 'untidy lives'. So by way of conclusion – and of comparison with Paul as precursor – I want to look briefly at the famous conversion scene in Book VIII of the *Confessions*, in the garden in Milan.

'From a hidden depth,' Augustine writes, 'a profound self-examination had dredged up a heap of all my misery . . . I felt my past have a grip on me.' Paul, we should remember, was not troubled, except by his righteous and angry wish to punish Jesus's followers. Augustine has gone into the garden of the house he is sharing in Milan, in a very troubled state. He is with his friend Alypius, whom he moves away from so he can weep, 'to ensure that even his presence put no inhibition upon me'. It is release and relief he is seeking from a great inner turbulence and strain. 'I threw myself down somehow under a certain fig-tree, and let my tears flow freely.' The fig tree is often taken to be a symbolic reference to Adam's fig tree in Genesis; and like Paul there is a falling to the ground, a loss of sight (this time through copious weeping), and the hearing of a voice. As with Paul it is something heard rather than seen that grips him:

> . . . all at once I heard the sing-song voice of a child in a nearby house. Whether it was the voice of a boy or a girl I cannot say, but again and again it repeated the refrain, 'Pick up and read, pick up and read' . . . I stemmed my flood of tears and stood up, telling myself this could only be a divine command to open my book of scripture . . . In silence I read the first passage on which my eyes lit [from Paul's Epistles], 'Not in revelling or drunkenness, not in lust and wantonness, not in quarrels and rivalries. Rather, arm yourselves with Lord Jesus Christ; spend no more thought on nature and nature's appetites.' I neither needed nor wished to read further.

Augustine overhears something from an unknown source linked with childhood, 'Pick up and read, pick up and read'; he can't remember, he writes elsewhere, whether this had a source he knew, but he does then remember someone else's reading and hearing a conversion experience, which prompts him to pick up the text by Paul, and apparently at random – reading 'the first passage on which my eyes lit' – he finds exactly the text he needs. One thing leads to another, but it is an experience complete in itself: 'I neither needed nor wished to read further.' It is as though a whole series of things suddenly come together. There is despair, an overhearing, a memory of a reading, and a reading experience; it is indeed a heartening transformation scene for departments of literature. For both Paul and Augustine it is, in part, a listening cure; and it is definitively a cure by language – or, as they would say, redemption through the word. Unlike Paul, Augustine makes no explicit demand; and yet, like Paul, he is now informed, so to speak, of what he really needs and therefore wants. In a more psychoanalytic vocabulary we can say that what has been revealed to Augustine is what we might call his desire. Knowing what he truly wants gives him a life; and, retrospectively, enables him to understand his past life (conversion experiences always explain, always make sense of the past; as though only a new future can make the past intelligible). A life, to be a life – in this tradition – requires an object of desire. A life is organized around the turning away from and the turning towards the object of desire (and in one version of this story it is as though the real object of desire is always already known; there to be re-found: Augustine's conversion was, in Lane Fox's words, a 'reverting'). Change is always the changing of the object of desire, or its recovery. The convert has finally got his wanting right, and so his life is in order. In this story, lives potentially have a real or true shape that can be – and have been – disfigured (there is a real and true object of desire). Conversion experiences reassure us that there is an essential life that one could be leading; and in secular terms this all too easily becomes the preferred life, or the missing life, the life missed out on. It is, that is to say, like many reassurances, its own kind of tyranny.

Either you do or you don't know what you want, and whether you do or don't it is not in your hands (no wanting comes with a guarantee); indeed, after Paul and Augustine, one might say, the question is not whether you want change but whether change wants you (you are in God's hands), or, more ordinarily, whether you want change (it is a shibboleth of psychoanalytic treatment that the patient comes to psychoanalysis to change by remaining the same). And so the examples of Paul and Augustine are hard to redescribe in secular terms; it is difficult to, as it were, convert the Judaeo-Christian conversion stories into their secular equivalents. If you take God out of the picture – a figure who desires you to change, and desires your desire to change; a figure who both knows what's going on, and knows what should be going on – there is a real sense in which you no longer know how to talk about change; except, of course, in the language of politics and philosophy and biology and psychology. Change, that is to say, is wedded to purpose; so the question becomes: change to what end? Clearly all of our stories about change are dictated by our cultural ideals; by the kinds of people we are being encouraged to be.

What is striking is just how much modern stories about change – represented here by modern commentaries on ancient conversion experiences – privilege evolution over revolution, incremental development over sudden radical transformation. There is a prevailing scepticism about the truth and the value of benign, dramatic personal transformation matched, so to speak, by a commitment to the malign, dramatic personal transformation of trauma; epiphanies have become objects of suspicion. As if lives cannot suddenly and startlingly change for the better, but they can suddenly and shockingly change, sometimes irreversibly, for the worse. As if the only conversion experiences that many of us believe in now are traumas. A loss of confidence, in other words, that there are or can be good forces for change; or a difficulty in locating such forces in the absence of a God. And that must be a sign of the times. Augustine's conversions, Robin Lane Fox writes, 'are not conversions to Christianity from non-Christian belief. They have emerged as conversions away from rhetoric, worldly ambition and sex.' It may be that the perennial question is: what do we want to be converted away from? And can conversion really do the trick?

Converting Politics

I

Everything looks permanent until its secret is known.

Ralph Waldo Emerson, *Circles*

Freud uses the idea of conversion as one of his preferred analogies to describe unconscious processes; conversion, in his account, is something we are doing unconsciously to parts of ourselves in the service of (psychic) survival. We convert, in his account, sexual excitement into somatic symptoms: through what he calls 'dream-work' we convert – at night, when we are asleep – perceptions, desires, 'latent thoughts' into 'the manifest content of the dream'. But, of course, he takes his analogy from what is often – in its religious and political versions – a determined and apparently all too conscious project. So it is of interest when the term is taken up politically, but with its Freudian associations and connotations in play.

Étienne Balibar, the French political philosopher and one-time student of the radical Marxist philosopher Althusser (who infamously murdered his wife), begins

73

his Wellek Lectures, entitled *Violence and Civility*, by considering what he calls 'the conversion of violence into institution, law or power/authority, and the possibility that there exists an inconvertible form of violence called cruelty'. By conversion he means 'a sublimation or spiritualization, but . . . also, and above all, a transformation of violence into (historically) *productive force*, an abolition of violence as a destructive force, and a recreation of it as the internal energy'. Politics, in Balibar's account, means describing or recognizing what might make politics impossible; and this depends upon a distinction between convertible and inconvertible violence. In Balibar's sense of conversion there is an abolition – of violence, as destructive – which is then re-created as a productive internal energy; violence is abolished as one thing (destructive) so it can be re-created as something else. If this was a simple question of pragmatism and language we could say: destructive violence is redescribed as productive forceful energy, and can then be used differently (for, say, civility). But this, clearly, is not what Balibar has in mind; and this is reflected in just how difficult it is for him to give an account of precisely how such a fundamental conversion takes place; the conversion by which a version of human nature is converted to the collaboration that is politics, in which people make a history worth having, and that justifies such conversion as is possible. Conversion in Balibar's account is a re-creation; and here conversion includes the belief that the already existing material, violence, can both be transformed, and transformed into something more desirable. This, at least, is his question: is this possible, and in what way is it merely wishful?

So much depends on what can and can't be converted; and on what is there, supposedly, to convert. It is not, of course, in Balibar's account, violence that is being converted, but people's violence (unless, that is, we believe there is something inside people called violence that could be chemically converted into something else). And we need to notice this because in conversion narratives something about a person gets selected out for conversion – sinfulness, violence, sexuality, religious belief, hearts and minds – while there are other things about people it wouldn't occur to us to convert (their dependence as infants, the colour of their eyes, their dead bodies). Conversion, that is to say, is a very selective form of attention (there is more to the convert than his conversion, whether he likes it or not). And it has to define what it converts in a certain way in order to make it convertible; in order to make us think it is subject to conversion (the Catholic and the communist both believe the material can be worked, the material of human nature; that a person can be re-created, or reborn, as something quite different). In this sense conversion is as much a defining process – however tacit, however implicit – as a transformative one. If something is redescribable in a certain way, you can convert it; if not, not. You can't convert a man into a horse, but you may be able to convert a man into a monster.

Just as Freud was wondering, to begin with – and Balibar's use of the word 'sublimation' and his references to Lacan keep psychoanalysis in mind, just as 'spiritual-ization' keeps religion in mind – what it was about sexuality that was convertible, for better and for worse, the founding force that Balibar is working with is vio-lence. And just as Freud was wondering what was intractable about sexuality and what could be satisfying about it, Balibar is asking us to wonder: if violence is the material of politics – the violence created by scarcity and frustration, by inequality and censorship, by exploitation and domination – what kind of civility might be pos-sible? 'To assert', he writes, 'that politics is a conversion of violence . . . is a way of saying that violence must prove convertible: it must be proven that it can be converted by politics, and that history is the process of this conversion'; as opposed, presumably, to history being precisely the ongoing failure of this essential conversion. The proof of politics, the proof that there is such a thing as politics, is that it can productively convert destructive violence. What Balibar calls 'the precariousness of politics' depends on such conversions. What he seeks to understand is 'how to "civilize" a revolutionary movement from within, how to introduce the anti-violence that [he] calls civility into the very heart of the violence of a social transformation'. Belief in conversion, that is to say, is inherently optimistic, if not actually utopian, in that it believes that the material to be converted – call it human nature – is essentially and potentially all to the good. All social transformation is violent and about violence, Balibar suggests, but it is only of value if it also has within it the civility of violence converted. If you believe, as he puts it, that 'a collective political movement' can 'transform structures of domination that will not disappear spontaneously'; or in a movement that is 'intent on changing change', as he puts it in an engaging phrase, then you must believe in conversion, and in what is there to be converted. Conversion, then, as the key to political transformation. Though the idea of conversion – even in Balibar's political version – does not seem to be intent on changing change, but on recruiting a rather too familiar form of change. Indeed, Balibar's doubts about his political project may be located – perhaps unconsciously – in his use of this particular word.

As always, then, to put it schematically, when it comes to conversion there are four questions: what is to be converted, who or what is going to do the converting, how is the converting to be done, and what is the intended (or presumed) result? For Balibar, violence is to be converted by politics: this converting is to be done through revolution or democratic process, but in order to create a politics of civility rather than an anti-politics of cruelty. To make a conversion story intelligible, however – and I am interested here in when, and why, 'conversion' is the word that comes to mind in any given context – we have to note what is deemed to be inconvertible: both what the inconvertible is supposed to be, and what its effects are (what might be left, or left over, after the successful conversion has taken place, and what might be its effects).

For the early Freud, what is striking about sexuality is just how convertible it is, into an array of symptoms and sublimations; and it is so convertible in his view because it is so disturbing, so unmanageable that it has to be managed through conversion (as if to say, what else can be done with 'it' but to convert it); the unbearable has to be converted in the service of adaptation.

'The incompatible idea', Freud writes in *The Neuro-Psychoses of Defence* of 1894 – and the incompatible idea is a sexual one – 'is rendered innocuous by its sum of excitation being transformed into something somatic'; it is for this that he proposes the name of 'conversion', in his first use of the term. Something incompatible is rendered innocuous by conversion. The individual has to survive her sexuality; so sexuality has to be convertible (that sexuality is the kind of thing that can be converted Freud, then, takes for granted, which itself opens up new ways of talking about sexuality: like an artistic medium it can be transformed, like a force it can be channelled, like a set of beliefs it can be redescribed; ideas can be transformed into bodily suffering). Conversion works, in Freud's account, by estranging us from our desire, from our real enjoyment, but with other pleasures.

Conversion for Freud is the best picture of the worst kind of adaptation: the sacrificing of sexual pleasure in the service of psychic survival. Psychoanalysis wanted to be the cure for the unconscious conversion of sexuality into symptomatology. There are, as we can see, the uses of conversion; and conversion as itself a problem. And then there is the problem of what is taken to be inconvertible, beyond that form of acculturation. The risk, for example, is that the inconvertible gives the lie to the convertible; that cruelty gives the lie to civility. That the key, as it were, to conversion is whatever it is in someone that resists conversion. Conversion after all may change everything by keeping everything the same. After Freud's incompatible idea has been transformed into a somatic symptom, it is still there; it has just been 're-created', to use Balibar's term, as a somatic symptom. God made the world out of nothing, Paul Valéry said, but the nothing shows through. In Balibar's civility, the destructive violence might make its presence felt even in its apparent absence. What has been converted is never abolished. This, at least, is Balibar's fear, and Freud's paradoxical hope.

What is inconvertible about sexuality, for Freud, is the desire for satisfaction; however sublimated, repressed, displaced or negated the sexual drive is, satisfaction is always being sought. And what is inconvertible for the late Freud is the death instinct, which can, by definition, be neither abolished nor re-created as something more enlivening and productive. For Balibar, what he calls cruelty – which is clearly related to Freud's death instinct, and more explicitly to Lacan's 'thing' – is the violence that can't be converted (i.e. controlled). And he cites as examples the geno-cides and ethnic cleansings of fascism. In both Freud and Balibar's accounts, the conversions that are pos-sible are haunted and informed by whatever is assumed to be inconvertible. Indeed the inconvertible is endur-ingly the potential saboteur of the converted; it is the conversion horizon. Or rather, the inconvertible is what sets an absolute limit to conversion (sex in Freud's account can be converted into symptoms, but then, as Freud famously remarked, the patient's symptoms *are* his sexual life). And by setting such limits, the incon-vertible exposes, or overexposes, the assumptions of those bent on converting people; assumptions about what they take human nature to be, or to be like. Assumptions about what might be called the purity of change; of how absolute and irreversible processes and projects of transformation can be (both politically, and personally). When Freud writes of the return of the repressed he is writing about conversions that are no longer sustainable; of conversion experiences that break down. People are also what you can't make of them, or make them into. We are, too, the conversions

we have resisted; and indeed the conversions we are
tempted by (and our repertoire of conversions avoided
will be telling us something about ourselves). In our
descriptions of ourselves it is never clear whether we
prefer to define ourselves by what we deem to be inconvertible, or by what can be converted; essentialism
being on the side of the inconvertible, pragmatism taking the supposedly inconvertible to be a tyranny and a
provocation.

For Balibar, whatever civility may be possible in
politics is always under threat from an indelible
cruelty – 'How does one keep from being fascist, even
(especially) when one believes oneself to be a revolutionary militant?' writes Foucault in his preface to
Gilles Deleuze and Félix Guattari's *Anti-Oedipus*. 'How
do we rid our speech and our acts, our hearts and our
pleasures, of fascism?' In the conversion experience of
growing up, we are clearly somewhere either unsuc-
cessfully converted fascists, or incompletely converted
fascists. It is a question, as Balibar puts it, of where we
'draw the borderline between what is convertible in
power relations and what is not'. Where we draw the
borderline, but also what happens to the inconvertible,
given, by definition, that we can do nothing, or very
little, with it (it may be that all we can do is acknowledge it and act accordingly; or assume, like pragmatists,
that we merely haven't had our way with it yet).

In all conversion stories there is, then, an abiding

fear – that can also sometimes be a reassurance – of what is inconvertible; of what resists the conversion of what is ultimately redescription, however hard-won. In Freud and Balibar's story, what is exempt from conversion is what we are endangered by; is, indeed, what threatens to make a mockery of all our conversions, of all our ideals (though in psychoanalytic language now we can say that we are sustained only by what is beyond our omnipotence: the worst thing that can happen to a child is that he succeeds in converting his parents). For Freud, the ultimate emblem of the inconvertible was, as I say, what he called the death drive, Thanatos. Though attempts are continually being made, no one can be converted to not dying, or not fantasizing about killing people. And, indeed, there hasn't, historically, been too much success in the project of diminishing, let alone converting, human violence. In this sense the will to conversion is always an acknowledgement of something intractable, or feared to be so. Where there is a conversion experience there has always been some kind of prior acknowledgement that there was something endangering that was unbearable; something deemed to be essential tempted and tempered by the hope of its transformation. Conversion, then, becomes the cure for a malign essentialism.

But if there is always the underlying fear (and comfort) of what about ourselves can't be changed, there is also the complementary fear (and desire) that anything and everything about ourselves can be changed; and changed by conversion, or something akin to it. The utopian political schemes of the last century – fascism and communism – were both, in their different and similar ways, committed to reinventing the human; to making completely new kinds of people; to re-creating human nature. Here there is a different kind of essentialism in play: at its most extreme, the essentialism of there being nothing essentially fixed about human nature; or in the case of fascism and communism, the recovery of a lost or abandoned human nature, a human nature perverted by capitalism, and racial and ethnic differences, that can be converted, through violence, to its putatively better and redeemed versions. Indeed, as Balibar knows, wherever conversion is spoken of, even the conversion of violence, violence is often required. And this inevitably complicates his project of converting destructive violence into politics, and into a politics of civility that he promotes.

Where Balibar, as a consequence, sees our inconvertible cruelty as always about to undermine our politics, and our confidence in politics, Wendy Brown sees neoliberalism – which she defines in *Undoing the Demos* as 'a peculiar form of reason that configures all aspects of existence in economic terms' – as 'quietly undoing basic elements of democracy' with its own subtle and not so subtle violence. Like Balibar, she has recourse to the idea of conversion, but in its more malign version, reminding us of the dangers of conversion in political life. It is, she suggests, a kind of contemporary conversion, that is undoing the possibility of politics, which she thinks of as democratic politics and refers to as 'the sovereignty of the people'. There has been what she calls 'the conversion of political processes, subjects, categories, and principles to economic ones'; and now that speech has become, as a consequence of this conversion, 'the capital of the electoral market, then speech will necessarily share capital's attributes: it appreciates through calculated investment, and it advances the position of its bearer or owner'. Clearly something about conversion experiences can tell us something essential about what is, and isn't, happening to us now politically. And by the same token, about what is happening to our language, our speech. Perhaps we need to remember that whatever the violence in conversion, language is always required. One is converted to, and by, something that can be spoken or written.

This conversion that Brown describes is a conversion of language; our speech becomes something we think of ourselves as investing in, and as simply and solely instrumental in our economic success; it serves our ambitions for personal gain. And so what we might have thought of as a shared language is a language calculated to make the shared itself a profit-making enterprise (the political being the official shared world). If language itself is a commodity – something to invest in – what will it then be unable, or unwilling, to do? (What will the poems of neoliberalism sound like?) If it is always the ambition of conversion (and politics) to create or force consensus, or like-mindedness, then it is always going to be the enemy of diversities and pluralism (conversion circumscribes a vocabulary, the right words in the right order). So for Brown, though she doesn't say this explicitly, conversion experience is the enemy of democracy; as though democracy depends upon delegitimating conversion as a good model for change, both personally and politically. Democracy fails when it allows conversion experiences to enter the realm of politics. As though where there was or has been conversion, there should be ongoing conversation and argument. A conversion is a conversation that has failed; it contains too many similar views. A conversion is a conversation that has ended. There are fewer rival claims to conciliate, or, indeed, imagine (we could think of the wish to convert as the wish to abolish rivalry; the wish to not hear too many voices). Her argument, as Brown writes,

> is not merely that markets and money are corrupting or degrading democracy, that political institutions and outcomes are increasingly dominated by finance and corporate capital, or that democracy is being replaced by plutocracy – rule by and for the rich. Rather, neoliberal reason, ubiquitous today in statecraft and

the workplace, in jurisprudence, education, culture, and a vast range of quotidian activity, is converting the distinctly political character, meaning and operation of democracy's constituent elements into economic ones. Liberal democratic institutions, practices and habits may not survive this conversion.

If this is a conversion – of democracy into plutocracy, of reason into neoliberal reason, of morals into money – what does a conversion entail, or involve, such that it can be this powerful? Brown intimates that this conversion experience is akin to replacing one language with another, or learning a new language, but a language without redress or dissent; a language that makes competing or alternative languages seem unrealistic, or not really languages at all; a language, indeed, that corrupts and degrades democracy. The opposite, that is to say, to the ways in which language has been described in so-called liberal education. A new kind of language that is an undermining of an old one: propaganda masquerading as a new realism. So in Brown's account, neoliberal reason is not a translation – which might be one alternative to conversion – but a violation. What she is actually describing is the final conversion, the conversion that puts a stop to all possible future conversions. And there is, of course, a sense in which this would be true of all conversions; the converted don't, by definition, assume that there will be future conversions. People don't tend to be converted to something for the next few minutes, any more than they can be colonized for a couple of days. Conversion proposes the kind of change that makes other or future forms of change unthinkable. It is the kind of change that renders all future change redundant. It quite literally gets you to the best of all possible worlds; and then stops time. And so, as Brown says in her lucid and compelling book, one inevitable casualty of this conversion to neoliberal reason, to this language of profitability and the market, is liberal education. 'The most dire entailments pertain', she writes,

> to the effects on democratic citizenship of this conversion in the purpose, organization and content of public higher education [she is writing of the United States, but also more generally]. After more than half a century of public higher education construed and funded as a medium for egalitarianism and social mobility and as a means of achieving a broadly educated democracy, as well as for providing depth and enrichment to individuality, public higher education, like much else in neoliberal orders, is increasingly structured to entrench rather than redress class trajectories. As it devotes itself to enhancing the value of human capital, it now abjures the project of producing a public readied for participation in popular sovereignty.

What Brown opposes to this neoliberal conversion of education – what she calls 'the saturation of higher education by market rationality [which has] converted higher education from a social and public good to a personal investment in individual futures, futures construed mainly in terms of earning capacity' – is egalitarianism, social mobility, broadly educated democracy, deep and enriched individuality, redressing of class differences, social and public good. This, then, is conversion experience as violent, imperial and, to use Balibar's term, cruel; it abolishes and replaces, rather than being challenging or enhancing; conversion as the alternative to conflict. Democratic reason, egalitarian reason, liberally educated reason, it seems, is all too convertible to neoliberal reason. What made it so vulnerable, or so desirous of such a conversion? What were the preconditions – another key word in thinking about conversion – that made this conversion both desired and apparently so successful? What we want from a conversion experience – and that we might want a conversion experience for rather than the many other forms of change available – must be a clue about what we feel to be lacking; about the experiences we aren't having in the experiences we are having. And about what it is about change that we can't bear (that it is always provisional and uncertain, say; that it is a risk that can never be guaranteed; that it has consequences we can't account for; that it makes us too vulnerable, and so on). Conversion is supposedly transformation with an insurance policy.

Converting Politics

That political theorists like Balibar and Brown should have recourse to a term of more than obvious religious connotations to explain what are for them never less than secular political predicaments is, I think, of some significance. As though there is something we can't adequately describe – at least as yet – without this religious vocabulary and its history. As though there are certain kinds of transformation – both desired and feared – that only conversion can account for, is the best analogy for; or that can provoke the most illuminating associations. As though the whole notion of conversion is a way of talking about extremity of need, and of fear (how we want to change and be changed is telling us something about what we think needs to be changed; and what is there to be changed). Or as though conversion needs to be backed by a deity – by an omnipotent and omniscient figure – to be legitimated. Wherever there is conversion someone claims to know what they are doing, and to know what is best for other people. But to know in the more absolute senses of knowing. Conversion experiences must, by definition, dispel scepticism. Ultimately it is doubt that has to be converted, the doubt about what one is being converted to; and doubts are to be converted not into more doubts – as they may be in liberal education – but into certainties. Conversion, we can say, betrays a terror of scepticism. One can only be converted to a cult.

Both Balibar and Brown, explicitly and implicitly, assume in one way or another a desire for conversion;

in Balibar's case an active political project of converting destructive violence into some form of civility, despite the connotations that the word entails. And in Brown's case it takes the form of an implicit question: what are people – liberally educated people, and others – wanting in being converted to neoliberal reason? What does it give them, what does it free them from, that makes them, us, ripe for conversion? So we need to ask the question – that almost by definition no one poses consciously to themselves – why might we want to be converted, to neoliberal reason, or to anything else? Why might that be our preferred form of change, or of changing things? If we didn't want to convert what Balibar calls destructive violence, what else might we do with it? If conversion can be at once a solution and a problem – as it is for Balibar and Brown – what kind of work are we using it to do? I want to say that the idea of conversion – the word itself, as Balibar and Brown illustrate – has stories to tell us about preferred and feared forms of change and transformation; and stories, most starkly, to tell us about the moral evaluation of change, about what kinds of change we prefer, and why. And about our fears of where our scepticism might take us; or where it might take us when there are no brakes on it. As though we assume there is either conversion or chaos.

If we want a representation of the perplexities posed by the familiar word, we need only look in Johnson's

great eighteenth-century dictionary – in which he
defines 'to profit' as 'to gain advantage', and a doubter
as 'one who hangs in uncertainty' – and read his definitions of the verb 'to convert' in the order in which he
presents them: '1. To change into another substance; to
transmute. 2. To change from one religion to another.
3. To turn from a bad to a good life. 4. To turn toward
any point. 5. To apply to any use; to appropriate. 6. To
change one proposition into another, so that what was
the subject of the first, becomes the predicate of the second (e.g. All sin is a transgression of the law; but every
transgression of the law is sin . . .).' Two neutral statements climax in the third, 'To turn from a bad to a good
life', which is then followed by three further neutral
statements, though the citation offered for the final
one is about sin. And this is followed by two neutral
definitions: 'To Convert: to undergo a change, to be
transmuted' and 'A Convert: a person converted from
one opinion or one practice to another' (with, it should
be noted, no explicit mention of religion). 'Conversion'
is simply a word for change – albeit of a sometimes radical kind (for 'transmute', Johnson has 'To change from
one nature or substance to another') – until, that is, it is
change for the better, 'To turn from a bad to a good life'.
It is entirely descriptive until it is dogmatically evalu-
ative. For us, now, I would say, it is unequivocally
morally charged, at least when it is used about people
(and it would be interesting to speculate why that might

be). People are only converted to what they take to be
the best life, or to what their converters take to be the
best life. And once they have been converted successfully,
they realize, in retrospect, that it was this conversion
that they were wanting. So their conversion gives them
a past and a future; it discloses what they had previously
been deprived of, and had been wanting – a different
nature or substance – though often without realizing it.
And it gives form to the future.

So we need to turn now to what a desire for conversion – however retrospective, however reconstructed –
is a desire for. Or, to refer back to Wendy Brown's
account, we may wonder, once we have been converted
to neoliberal reason – when we no longer talk about the
good and why we value what we value, but only talk
about (and in terms of) profitability and money – what
we will make of our former selves, and their supposedly
abundant preoccupations, which were, it turned out,
never what we really wanted; which were, in fact, a sign
of our unacknowledged impoverishment, a misrecognition of our nature. Unless, of course, there is something
inconvertible about us and we end up either as double
agents, half converted to neoliberal reason and half
violently opposed to it; or entirely and so probably very
violently opposed to it, compelled to find oppositional
forms of persuasion that are not conversion by other
means; if we are, that is, in Balibar's useful phrase,
'intent on changing change'. As Balibar says, violence

produces counter-violence; 'counter-violence', he writes, 'presents itself as second and as such a legitimate reaction to a "first violence" generally presented as illegitimate', and as we can say, conversion tends to produce counter-conversion, which is sometimes reactive to what is understood to be an initial illegitimate conversion. We don't tend to think of conversion as reactive to prior conversion, and perhaps we should, especially if we are to consider the alternatives to conversion, if there are any; that is, our other preferred forms of changing each other and being changed, and the reasons we can then give for these preferences. And after, of course, a long and troubled history of conversion experiences. Civility, we might say after Balibar, being an example of the conversion of violence, or as a disguised form of what Balibar calls cruelty; the converted always telling a story of purity and danger, of what, and sometimes who, needs to be got rid of in the sustaining of faith. The converted are nothing if not paranoid; they always know who the enemy is, and usually what is to be done with them. Conversion as another word for scapegoating.

Conversion, in other words, also invites us to talk about how we would like to influence and change each other (and ourselves) if we are not seeking to in any way convert each other (we might, for example, think or have thought of a liberal arts education as an alternative to conversion experience; or even the sometimes liberal art of psychoanalysis). It also invites us to consider how we imagine, how we picture, how we describe, significant changes happening to ourselves and other people. And not necessarily or only with a view to arranging more efficiently the changes we seek, as we must in political life; but also with a view to allowing and accounting for the unwanted or surprising changes that we might only in retrospect value, or be puzzled by, or not value at all (at one end of this spectrum there is the traumatic and at the other end the enlivening surprise; conversion is towards the traumatic end of the spectrum). Because conversion, as a way of changing people, has explicit aims and purposes, however covert or unintended its intentions turn out to be. Conversion is the change that assumes it knows what it is doing, that always already knows its own value; or rather, the converter always assumes, indeed must assume, that he knows what he is doing; he is convinced of his own virtue. So to be converted to something is, as I say, to have entrusted oneself to an omniscient figure, a figure who knows what's best for you, which is a form of omniscient knowing (to know what's best for someone presumes an inordinate knowledge of them and of oneself: a knowledge sometimes useful in child-rearing, but otherwise suspect). The worst thing we can do to goodness is to be too convinced of our own.

So in considering alternatives to conversion, we are imagining the paradoxical idea of wanting to change ourselves and others, but in unpredicted and unpredictable

94

ways; ways that can be evaluated only prospectively and retrospectively, but never finally or definitively; in which the only evaluation is an ongoing evaluation (so, for example, 'outcome' would be a meaningless term). It is to imagine changing people, but using known aims and objectives as at best means rather than ends, and at worst as decoys; so, for example, every so-called cure would be just one stage of what could only accurately be described as an experiment in which the criteria of success would always be evolving, a transition to unknown future destinations (aims would not be targets; we would want to be cured of our cures). So perhaps the opposite of conversion is experiment; traditionally, the experimentation of empirical science in its conflict with religious revelation, but more to my purposes John Cage's definition of an experiment in his book *Silence*. 'The word "experimental" is apt,' he writes, 'providing it is understood not as descriptive of an act to be later judged in terms of success and failure, but simply as an act the outcome of which is unknown.' Conversion depends upon a supposedly known outcome. Alternatives to conversion, that is to say, are a way of talking about a wish to change uninformed by a known outcome, or teleology (one way of clarifying this would be to say that you wouldn't want to go to a medical doctor who had no idea what a cure was, but you might want to go to a psychoanalyst who wasn't too sure what a cure was). It would not idealize change at the cost of its aims and objectives; but it would include in its description of intended change the simple and infinitely complicated idea of unpredictable consequences, of the future as something that can be wished for but can't be engineered. Targets that must be missed in order to be met.

Balibar's 'civility' and Brown's different 'liberal democracy' both raise the question of how one might have strongly held values without, as it were, turning into another version of the enemy in trying to sustain them; how it would be all too easy to become uncivil and cruel in defending civility, or too instrumental and self-serving in the defence of liberal democracy and the education it fosters. It is surprisingly difficult to avoid becoming one's own worst enemy; to avoid fighting fire with fire, or becoming symmetrical with whatever one contends with (bullying the bullies, and so on). So it would be easy to believe that conversion, then, was the only cure, so to speak, for conversion. Or perhaps the whole notion of conversion – its practice and its process, its premises and its promises – can give us important clues about what else is to be done. Indeed, a clue about what might be called non-paranoid forms of change, forms of change that are not organized around identifying the enemy (change without scapegoating, without good riddance). Change in which there is what the French psychoanalyst François Roustang calls 'a sense of distinction that can in no way be transformed into mastery'.

Emerson's word for an alternative to conversion is 'provocation'; Freud's alternative is 'free association'. Both are ways of transforming oneself and others, without mastery, and by relinquishing mastery (as though mastery itself is something we have been converted to, is a resistance to change, and so can only be part of the problem). And for both Freud – as we saw earlier – and Emerson, 'conversion' was also a key word in their writing. Freud, I take it, was one of Balibar's provocations, via Lacan (and Althusser), while Emerson was one of Brown's significant precursors in the quest for an unenslaved, non-exploitative, new American democracy.

II

Every ultimate fact is only the first of a new series.
Ralph Waldo Emerson, *Circles*

Freud assumes that the patient – the psychoanalytic patient – has already undergone his conversion experiences. Indeed, that conversion is virtually a synonym for acculturation and adaptation. If he has not literally, in Freud's words, developed hysterical conversion symptoms, he will have, according to Freud's psychoanalytic description of things, converted incompatible ideas into

something else – symptoms, displacements, sublimations, dreamable dreams – in an attempt to render them
as innocuous, as undisturbing as possible. Simply in the
process of growing up and making his desire compatible
with his psychic survival, the Freudian individual is
a convert: converted by acculturation to acculturation:
converted both to civilization and to being one of its discontents. Conversion as the false cure for ambivalence.

Freud's attempted cure for some of the suffering this
involves is what he calls 'the method of free association',
in which the patient – and the analyst, in a different
way, through her free-floating attention – suspends his
scepticism about his own thoughts and feelings, with a
view to speaking more freely. And this relinquishing of
certain judgement about one's thoughts and feelings –
this holding one's internal censorship at bay – reveals
the sheer scale of the patient's uncertainty about himself.
Where there was conviction – 'I am the kind of person
who . . .' – there are competing accounts; where there
were objects of desire, there are various objects of desire,
and, indeed, forms of desiring; where there was a self,
there is a new-found disarray that the word 'self' doesn't
seem to cover; where there were familiar memories,
there are screen memories; where there was a preferred
vocabulary, there are competing vocabularies. It is, one
might say – taking up Freud's vocabulary – a deconversion experience in which if one is converted to anything,
it is to the infinite complexity of one's own mind, to the

unending conflict between one's urgent desires (one is converted, in short, to an imaginative scepticism about conversion itself). And therefore an awareness of why and how one might be, and always have been, tempted to narrow one's mind, to simplify oneself, and partly because of the sheer difficulty of containing multitudes; multitudes, that is, of desires, and griefs, and conflicts, and beliefs, and pleasures. Because, in Freud's view, desire is always in excess of the object's capacity to satisfy it, desire enlivens us by endangering us. Conversion becomes, in this account, a wished-for narrowing (and so securing) of the self, but disguised as its true revelation. The defended self displaces its counterpart. Psychoanalysis at its best, then, might cure people of their wish for conversion; of their wish both to convert and be converted. Or perhaps, a little more realistically, psychoanalysis invites us to imagine what the unconverted life might be like. Or what we might be left with after the deconversion experience of free association.

In his 'Two Encyclopaedia Articles' of 1922, in the section entitled 'The Fundamental Technical Rule', Freud spelled out what he called the 'procedure of free association':

> The treatment is begun by the patient being required to put himself in the position of an attentive and dispassionate self-observer, merely to read off all the time the surface of his consciousness, and on the one

hand to make a duty of the most complete honesty, while on the other not to hold back any idea from communication, even if (1) he feels that it is too disagreeable or if (2) he judges that it is nonsensical or (3) too unimportant or (4) irrelevant to what is being looked for. It is uniformly found that precisely those ideas which provoke these last mentioned reactions are of particular value in discovering the forgotten material.

It is as though Freud is saying that conversion experiences instal a regime of censorship – an infinite set of incompatible and rejected thoughts – and so the psychoanalyst must attend both to what is censored, and to the forms the censorship takes. It is essential to the palpable design of the conversion experience to ensure the omission of all apparently disagreeable, nonsensical, unimportant and irrelevant ideas. It is the quiet, insidious violence of repression that Freud exposes; but without using anything akin to the language of politics, when he is manifestly describing an oppressive internal regime; a regime that can tell you – has told you – what is important, so you always already know. (How can you know what is important to you beforehand, before finding out? We can't help but note, that is to say, the omniscience of the censors implied in Freud's account: they are like mad aesthetes.) What the conversions of contemporary life exclude – and conversion is by definition selective and exclusive – the psychoanalytic treatment will re-include. And then the question becomes: what is to be done with the re-included ideas? Or, in Freud's language, once what he calls 'the forgotten material' has been discovered, what does the analyst make of it, what does he use it to do? The next section of Freud's 'Encyclopaedia Articles' is entitled 'Psychoanalysis as an Interpretative Art'.

After the deconversions of free association, is there to be a kind of reconversion, if only to psychoanalysis and its new method of treatment? What does the analyst do with the patient's free associations if he – and the patient – resists the temptation of a new conversion, a better conversion? Freud's psychoanalytic method, in other words, dramatizes a question about change and transformation – both personal and political – and a question about interpretation. How conservative are they, and what are they conserving? When we talk about conversion, we are always confronted with the alternatives to conversion, and what they sound like, especially once we have become sceptical of conversions; not least because they depend upon the censorship of scepticism. What the convert is converted to is severely circumscribed forms of permissible doubt. Is there only ever going to be, more or less, more of the same? 'Not only do I not think the structures of social domination – whether economic, cultural, or sexual – will dissolve on their own,' Balibar writes, 'I also do not think that their consequences can always be prevented from getting worse without violence, or without the becoming-violent of a social force that is the object of a form of repression that is itself violent. That is precisely why I consider it so important, and so relevant to the present, to understand – retrospectively, and therefore prospectively – how to "civilize" a revolutionary movement from within.'

For Balibar, real change is revolutionary change; violent repression, violent domination and exploitation, has created the need for violent revolution, but with the hope, in the middle, of some civility, of civilizing a revolutionary movement, as he puts it, 'from within'. Freud, on what might be called a smaller scale, is proposing, also, a restoration of order, and a more inclusive, less silencing regime. But we have to distinguish between organized and over-organized aims and objectives, for both the putative revolutionary and the psychoanalytic patient and his analyst. Both, of course, would claim that they knew what they wanted; or at least believed it possible to know what they wanted (less misery, for a start). 'The patient's associations', Freud writes in 'Psychoanalysis as an Interpretative Art', 'emerged like allusions, as it were, to one particular theme, and that it was only necessary for the physician to go a step further in order to guess the material which was concealed from the patient himself and to be able to communicate it to him.'

From the apparent disarray of free association – of the patient saying, uncensored, whatever comes into his mind – 'one particular theme' emerges; the patient allows himself to disrupt his themes and another theme emerges; by unsettling the old themes you get a new theme (a new theme, that is, of course, a very old, repressed one). But could there, for example, be something other than themes in a person's life (intensities, perhaps, or affinities)? And what use might they be? Because it seems as though psychoanalysis is helping the patient to get some better themes; that the analyst's and patient's conversion to themes – to a life being themed – is firmly in place. The patient is deemed to have got his themes wrong, not to be suffering from having themed himself, and allowed himself to be themed. Should we, then, be converted to the themeless life? How else might we take it up if this description happens to appeal to us? What would the analyst be hearing if he wasn't hearing themes; if he wasn't on the lookout for repressed themes? Is it, as Deleuze suggests, impossible to talk to a psychoanalyst because he always knows what you are talking about? We have to imagine what it would be like to talk to someone who didn't know what one was talking about, as a way of alleviating one's suffering; and why someone might want to do that. Talk to the unconverted.

Can you use an idea that you have not been, in some sense, converted to, and how would you do it? This, one could say, is the radical experiment that is psychoanalysis. So when Balibar wants to civilize a revolution from within, convert some of its violence into civility, is he wanting a newly unengaged, engaged politics? Freud is certainly intimating, I think – partly, as he might say, through negation, but also through free association that is met by an analyst's free-floating attention – the possibility of a relationship to oneself and others that is not based on conversion; a relationship in which persuasiveness is the problem not the point. And then we are left to wonder what a non-persuasive politics could be like.

'What is the good', Freud writes in his *Introductory Lectures on Psycho-Analysis* (1916–17), 'in the sphere of the intellect, of these sudden convictions, these lightning-like conversions, these instantaneous rejections?' He wants us, in what he calls the sphere of the intellect, to make the time and space to think and talk; and so to be sceptical of the immediacies, and the certainties, and the rejections that conversions entail. And, presumably, that revolutions entail.

Believe It or Not

I

All things are subject to time, moreover: they possess
no complete identity in themselves, but are always in
the process of becoming something else, and hence also
in the process of becoming nothing at all.

David Bentley Hart, *The Experience of God*

A thirty-six-year-old man tells me in a session about his
only ever having had sex with prostitutes. I ask, why
prostitutes? And he replies, 'Because prostitutes are the
only women my parents will never meet.' I say, 'And if
your parents did meet a woman you desired . . .' and he
interrupted me and said, 'They'd convert her, they'd
take her over, they'd kidnap her, she'd belong to them.'
He would be excluded, left out and robbed. The pleasure would be all theirs. He experienced his parents as so
omnipotent, so voracious in their presence – they could,
as he put it with terrified irony, 'be anywhere, you never
know' – that he had to go to great lengths to secure a
measure of privacy. And a measure of privacy was a

measure of desire; having the mental space, unimpinged upon by his parents, to be able to want. It is a picture of the parents as asset-strippers; anything he wants and values they must appropriate.

There was, of course, much more this man said about his parents, and indeed about his erotic life. But for the purposes of this chapter I want to single out a simple double act – the implacable converters (for this man, the tyrannical parents) and the helpless onlooker (the petrified or bemused voyeur of a conversion process). The implacable converters can feel like thieves, and the helpless onlookers might feel that they have witnessed a daylight robbery. And then there is – in the middle, as it were – the person being converted.

This scenario can be, and is, triangulated – there are the converters and the person being converted – and in these pages I (and we) take up the position of the witnesses of conversion. We are the more or less helpless onlookers, as my patient imagined he would be. And like this man, we may not be as safe as we might want to be. We may, for example, as onlookers, identify with both the converters and the converted; indeed, at its most extreme, the whole scene might send a depth charge into our own histories; our histories of having been fashioned and self-fashioned, changed and stultified, through our relations with others. We might be like people returning to the scene of the crime, which is sometimes the crime of people knowing what's best for us, or of our knowing what's best for them. Because in the conversion drama we are, in a sense, re-experiencing extreme versions of seduction and persuasion and pedagogy; of influencing and being influenced; of people changing – through a relationship with another person and their group – not merely their minds, but their lives. Of people breaking the spell of incremental, gradual biological change; what we might have taken to be an unfolding – in our descriptions of organic growth and development – becomes a rupture, a revolution, a superseding. We are seeing what it looks like, what it sounds like, what the participants imagine might be going on. And the 'it' is the extraordinary effect people can have on each other. This is where our fascination and our fear is: the effect of transformation.

Conversion is the exchange that demands change, and claims to know the change that is needed. How do we go about finding out the kinds of change we might want for ourselves? And how, if at all, can we disentangle these available forms of change from the changes other people are keen to persuade us of? Whatever else a culture is, it is a repertoire of desirable forms of change. And then, of course, we have to distinguish between desirable ends and desirable means: I may not want to be heterosexual – I may be, at its most minimal, ambivalent about my heterosexuality – but do I want to be converted to homosexuality? I may be interested in Islam, but I might believe that conversion to Islam might waylay or even sabotage my studying it. The other thing, that is to say, that we can single out from my patient's experience – and not only his experience, of course – is the fear of conversion as a process of transformation. It is notable that in my patient's fantasy of his parents converting or kidnapping any girlfriend of his, he assumes that she will have no agency, no powers to resist his parents; that is, he ascribes to her the inexorable effect of his parents' words on him. But whose parents' words have not been effective, if not quite that effective? The fear of conversion could be described, then, as a fear of losing one's mind; or of losing whatever it is one imagines can regulate and think about the effect of other people, and their words, on ourselves. Which is partly the appeal, the poignancy, of their need to be believed. This is one of the things that people do to each other, that we grew up having done to us and gradually doing ourselves, and which makes conversion so resonant for us: we try to make people believe us, and we try to find ways of managing when they do and when they don't.

And in the finding of representations of what people can do to (and for) each other – which is what, presumably, the so-called humanities, and psychoanalysis itself, are for – the conversion scene is akin to what psychoanalysts, after Freud, call the primal scene. Or rather, it might be useful to think of the conversion scene – so vividly imagined by my patient – as in some ways analogous to the primal scene, in which the child witnesses or imagines his parents' sexual relations. 'I have explained this anxiety', Freud writes in *The Inter-pretation of Dreams*, 'by arguing that what we are dealing with is a sexual excitation which [the child's] under-standing is unable to cope with and which they also, no doubt, repudiate because their parents are involved in it.' In describing and accounting for conversion experi-ences, we are like the child in Freud's primal scene, somewhat naïve and inexperienced: we are witnessing an intense exchange – usually a verbal intercourse – that absorbs the participants, and excludes everyone else; and which, to some extent, our understanding is unable to cope with (you are either in it, having the experience, or you are outside it, at an unfathomable distance). Freud's sentence is translated in a way that sustains the ambiguity of the sexual excitement; it could be both the child's, as voyeur, and the parents', as agents. And by the same token, we as the voyeurs of conversion experi-ences may be alternately envious and appalled, excited and critical, provoked and dismayed. The conversion scene, in other words, like the sexual scene, can leave us scrambling around for a satisfactory position, so to speak. It over-organizes and disrupts our attention. It is so unsettling that we need to come to conclusions about it. It stimulates and baffles a wish for knowledge, a running for cover. The primal scene, according to Jean Laplanche and Jean-Bertrand Pontalis, following Freud's relish for generalization, 'is generally inter-preted by the child as an act of violence on the part of the father'. The unconverted can think of conversion as an act of violence. And then we must wonder what we think is being violated. They can also think of conversion as robbery and loss, just as the child has been robbed of and lost his parents when they live their erotic lives together.

This man, I believed, identified with prostitutes because he thought of prostitutes as people who couldn't be appropriated (weren't converted by sex, could be bought but not owned). Prostitutes were people, in his mind, who had dissociated sex and attachment, and so could never suffer from loss. And he thought of parents as people who took and took over what belonged to their children (as people who couldn't be converted by their children, whom they owned without buying). He had located the wish to convert, take over, kidnap, own – an interesting list – in his parents. But what did this make him? His answer was 'A free agent', a person wanting neither to convert nor to be converted. I said, 'An agent free of what?' and he replied immediately, to his own slight bemusement, 'Free to be free'; and then qualified it by saying, 'Free to not have to worry about being free.' He wanted, he said, to be 'neither the drug nor the drug dealer'.

For my purposes this is a useful nexus of associations: conversion, invasion, ownership, addiction. This man, I take it, is wondering about the links between these things, as well he (and we) might; and is trying to work out an alternative, in human relations, to this set of elements, conversion, invasion, ownership, addiction. They are all forms of entrapment; and in his account they are exclusively malign. This reflects in interesting ways the moral dilemma created by conversion; because the converted, like their unconverted counterparts, are committed to what was once called the good. Or, to put it slightly differently, people only convert to what they believe matters most to them. In that sense the converted vindicate Socrates' belief that it is only the good that is sought. 'No one', Plato describes Socrates as saying in *Protagoras*, 'freely goes for bad things or things he believes to be bad; it is not, it seems to me, in human nature, to be prepared to go for what you think to be bad in preference to what is good. And when you are forced to choose one of two evils, nobody will choose the greater when he can have the lesser.' The convert is going for what they think of as good, and does not tend to think of himself as going for the lesser of two evils. In this version the good is irresistible, and yet there are competing goods. Everyone is in pursuit of the good but no one can agree about what the good is. In this view everyone has good intentions (we can say of the converter 'he meant well'). Those who convert people are promoting the good, their good; they are promoting virtue, their virtue.

And then Donald Winnicott, the twentieth-century English paediatrician and psychoanalyst of dissenting stock, can say, in a book appropriately entitled *Playing and Reality*, that 'madness is the need to be believed'. It is mad, then, to need other people to believe in your version of the good? What should one do with one's virtuousness? If we don't want to convert people to what we most value, does that mean we don't really value it, or that we do? Or does it mean that we don't value it more than we value other people and their dissenting views? If we don't need to be believed, what else should we, or could we, be needing from other people? This, as Winnicott would stress, is a question of dependence. And as conversion usually comes from outside – we don't tend to talk of people converting themselves: we think of conversion as coming out of a relationship with somebody else – it can usefully be construed in terms of dependence and its vicissitudes.

Winnicott is saying here, extraordinarily, that when we depend on other people to believe us it is a sign of our (temporary) madness; that is to say, a sign of something unintelligible and excessively disturbing, a sign of profound unhappiness and deprivation. Like my patient, he believes we are endangered by people who want to convert us (which in the first instance is our parents; or, in Winnicott's language, could be more accurately described as the mad part of our parents); and by our wish to convert and be converted (perhaps one of the ways we look after people, or placate them, is by believing them or persuading them; that faiths and cults are the mental hospitals of their leaders). Winnicott is, of course, talking obliquely about his fear of psychoanalysis – especially Kleinian psychoanalysis – as a conversion therapy. People who believe in conversion believe in intractable certainties; and with this comes the assumption that only certainties are dependable. People are only ever converted to something they believe they can depend on. Winnicott's striking formulation at least allows us to wonder how dependence and uncertainty can go together: clearly, our dependence on uncertainty – our dependence on our scepticism – is going to be quite different to our dependence on what we take to be guaranteed (God, nature, the leader, the ideology, psychoanalysis, the mother, and so on). For Winnicott, in other words, the developmental question for everyone is: how can I depend on someone whose reliability can never be guaranteed? It is a straight line from this to the idea of faith; and the equation between believing in and depending on.

Dependence is always an experiment in like-mindedness. So Winnicott can be taken to be asking, in his own psychoanalytic way, what are the preconditions for our need to be believed? And what do we fear will happen when we are not believed? Or, if we depend upon being believed, what are we depending on? Questions like this might help us to clarify the differences between conversion, addiction, entrapment and ownership; and whatever the alternatives could be in human relations.

Conversion, addiction, entrapment and ownership, we should note, are all forms of consistency; and if and when consistency is equated with reliability, or dependability, or trust, these will be alluring, if malign, options.

Winnicott proposes a capacity for surprise as an alternative to the need to be believed; an openness to surprise, a desire for it being integral, in his view, to a realistic and enlivening dependence on anything or anyone. In this view, whatever, or whoever, is available to be believed or believed in would be assessed according to their tendency to minimise surprise (this is what Charles Lamb was pointing out by writing, with hospitable ambiguity, that 'it is good to love the unknown'). Winnicott is proposing, in the Protestant way, that we should put our faith in what we cannot possibly know. As if to say – though it may be a question – you can't be converted to the unknown. Conversion would be the wrong word.

So being believed, for Winnicott, could be redescribed as being taken on your own terms; and the need to be believed, at its most extreme, is the need to live in a world without other people (without the dependence and exchange, and potential for play and uncertainty that he believes constitutes a good life). If you need to be believed, you need to live in a world of collusion and accomplices (for my patient, that would mean women as prostitutes). In a conversion experience the wish to change someone usurps the wish for exchange; the wish to make something specific happen pre-empts the possibility of surprise. The converted, one could say, are circumscribing their possibilities for surprise; averting what they take to be an irrelevant and distracting shock of the new. Conversion, then, is an attempt if not to dispense with scepticism, then at least to put it in its place. When Winnicott famously said that health is much more difficult to deal with than disease, I want him to be saying that being ill is like a conversion experience: it narrows our minds, it over-organizes our attention, it prescribes its own limits. In health we have nothing to protect us but our fear of our own freedom.

It is possible to see Winnicott's work, then, as an attempt – in psychoanalysis, and reactive to the history of psychoanalysis, and not only to that history – to think through the need to be believed; to find out what else, if anything, people could do together other than try to convert each other (the technical psychoanalytic term for conversion was 'identification'; the political term was 'imperialism'). Winnicott is working out what psychoanalysis could be if it wasn't a conversion experience; but using psychoanalytic language to raise questions outside its small domain. If William James's pragmatism encourages us to believe in something only when it suits us to – or only, in Richard Rorty's terms, when it helps us to get the lives we want – we can see Winnicott using psychoanalysis pragmatically, and implicitly encouraging us to do the same. That is, using it – and by implication using his own work, and everyone else's – only when and where it works for us. And not, therefore, as something anyone has been converted to, and has to keep reiterating (conversion guarantees and justifies its preferred repetitions). The problem with (and for) psychoanalysts has always been that they have had to believe in psychoanalysis. The problem for everyone is that they have needed to believe what certain people have told them (starting with their parents). What can we do with other people's words if we don't need to believe them; when that is a pressure we can resist? If, for some people, conversion (the need to be believed) has become the negative ideal of human exchange – what ordinary conversation tries to ward off, or keep at bay. By the same token, we need good descriptions of how people can change each other; we need to know something about the descriptions that have stuck: descriptions of, to put it as simply as possible, people having what some people think of as a good effect on each other, and of people having what some people think of as a bad effect.

I want to compare and contrast some well-known remarks by Socrates on education in Plato's *Republic*, and some less-well-known accounts of sexual perversion by the American psychoanalysts Arnold Cooper and Robert Stoller. Both these accounts need to say something about conversion to say what they want to say; and both implicitly make a simple but salutary point: that to be converted is to want to be like someone else. It is, in psychoanalytic language, a process of identification. It is to be turned away from something or someone, and turned towards something or someone else.

II

> In another world it is otherwise, but here below to live
> is to change, and to be perfect is to have changed often.
>
> J. H. Newman, *An Essay on the Development*
> *of Christian Doctrine*

In a compelling definition of what Rowan Williams suggests is 'terrible' about what he calls 'bad religion' – 'it's a way of teaching you to ignore what is real' – he offers us a question in the guise of an answer. In what he refers to as 'the healthy response' to religion, 'one of the tests of actual faith, as opposed to bad religion, is whether it stops you ignoring things'. Faith, he continues, 'is most fully itself and most fully life-giving when it opens your eyes and uncovers for you a world larger than you thought – and, of course, therefore, a world that's a bit more alarming than you ever thought. The test of true faith is how much more it lets you see, and how much it stops you denying, resisting, ignoring aspects of what is real.'

For the secular, Williams's definition of faith – it

uncovers a world larger than you thought, it stops you denying, resisting, ignoring aspects of what is real – could easily be applied to the study of literature, or to philosophy, or to psychoanalysis. Indeed, Williams gives as good a description as you might want of the aims of a liberal education; and it is surely telling that this comes as an account of good religion. Not many people now would claim to want anything that aimed to close their eyes, or that made the world smaller and narrower. And yet, of course, Williams reminds us why we might be always tempted to narrow our minds; a world larger than you thought is 'a world that's a bit more alarming than you ever thought' (the pastoral care is in the word, but also the pun, 'bit'). If a world larger than you thought is a bit alarming, then thinking is the way you regulate your fear.

Williams, in his account of faith and good religion in *What is Christianity?*, is not ostensibly trying to convert anyone; he is making a case for the value of Christianity. But what kind of people would we be if we denied the value and the values of the good religion he promotes? If we were in favour of ignoring things, denying and resisting aspects of what is real? He knows that it is not only or exclusively Christianity that protects these values; but we are challenged – as we may often be in persuasive, conducive writing – to wonder what we are doing, who we are being, when we resist conversion; or when we agree but don't think of ourselves as having been converted. Having one's eyes opened is after all a persistent and insistent figure for conversion. The question is: when your eyes have been opened, are you seeing more of what is real and true, or are you just seeing something else, something you didn't see before, but that is taken to be essential by a particular group of people? To put it crudely, are the converted better empiricists, or better believers? Can they, at long last, see what's there in front of them, or can they see what some people need them to see? And how, if at all, can they tell the difference?

It is instructive to describe converts – to religions, to ideologies, to ideas, to causes – as people who want to be like someone else, as like as they can possibly be. Were it to be formulated it would be as though people were saying to themselves, 'I want to be the kind of person who reads poetry, I want to be the kind of person who creates revolutions, I want to be the kind of person who can kill infidels and go to heaven.' When Williams writes of the gospel, he is writing of the potential Christian's wish to be – which at least in the first instance means to be like – a certain kind of person. 'That sense of revelation which invites you to change,' he writes, 'that's part of what the gospel is about, but it's even more, because the revelation is itself a revelation of an action of love into which you are invited to come, with which you are invited to cooperate. Come and see. Come and see whether it's possible for you to let go of that anxious and destructive self in the face of a promise of radiant beauty; to be made alive in this way.'

Again, we might say, we are being given a proposition – to love and not be anxious and destructive, to
experience radiant beauty – that it would seem churlish
to resist. But at its most minimal it is a simple proposal:
be as much like Jesus as you can be (Jesus is the picture
of your best self: your best self is your best aspiration).
Want what Jesus wants, not what you thought you
wanted (conversion is always the conversion of wanting).
So the question is not why do we want what other people
want – what else could we do, how else would we do our
wanting? – but who do we single out to be like, and why?
Why do we want what Mao or Beyoncé or Freud wants?
Why them? Freud says, in a memorable contribution to
all this, that we want to be what we cannot have. We
should add to this that because no one can ever actually
have someone else – actually incorporate them such that
they become them – there is only the being like. To begin
with, our parents, our families, are the objects of our
attention and desire (we organize our attention and desire
around them); and then gradually we cast our nets wider
(we like and want to be like people outside the family,
and this period when the child begins to find pleasure
outside the family is momentous). But this being like, this
inevitable process and project of identification – and I'm
taking identification, wanting to be like, as the ordinary
word for conversion – creates two interesting extremes
that I want briefly to consider.

There is the compulsion to identify, as a self-cure for

traumatic experience. This, one could say, if it wasn't a contradiction in terms, is conversion in the absence of alternatives, conversion as a life-or-death matter (identification with the aggressor – turning yourself into the person who harms you – is the emblematic example). But what psychoanalysis calls sexual perversion – which is itself an identification with an aggressor – is my instance here of the conversion born of apparent necessity; in a so-called perverse solution the child becomes, in his mind, like the mother who tantalizes him, in order to survive her tormenting care. This is identification – being like – as compulsive and essential for survival. And then there is the radical alternative, as it were, that I want to begin with: the wish to become someone no one could possibly be like, and who is supposedly not like anyone else (these two things can go together but they don't always). People who want to impress without the desire to convert, in which the possibility of conversion is a missing of the point. These people, one might say, offer us a lot, but nothing that we can be converted to. In fact they exemplify the sense in which conversion is a form of misrecognition; the wish or the desire to convert is a terrible misunderstanding of the problem conversion is expected to solve. Indeed, it is the problem masquerading as the solution; conversion as bad faith, as symptom, as self-avoidance through self-deception. The distinction, then, is between those who, consciously or unconsciously, encourage us to be like them, or force us to imitate them out of fear and intimidation; and those who inspire us, for want of a better phrase, to become ourselves, to become whoever else we might be. There is, then, our wish, or our need, to identify with others; and the pressure our chosen others may put on us to be like them.

III

What does the animal do? It reminds us that living and dying is the thing we must do.

Dorothea Lasky, *Animal*

In *The Art of Living*, the philosopher Alexander Nehamas describes Socrates as a new kind of model for Montaigne, Nietzsche and Foucault. For each of these philosophers Socrates was a positive and negative ideal; but also an essentially puzzling and paradoxical and therefore inspiring model. Partly, Nehamas suggests, because of what he calls Socrates' 'silence' – Socrates says virtually nothing about himself and left no texts of his own – but also partly because of the way Plato describes him as practising philosophy. Nehamas describes Socrates as effectively a philosopher without a method, but with a distinctive way of working that was simultaneously and inextricably a way of being himself.

Socrates' infamous irony and his anonymity – his resistance to, and enquiry into, laying down the law – are a
strange provocation (he isn't Christ, his guidelines about
how we should live are relatively unrestrictive). He promotes the project of the pursuit of virtue and reason by
exposing his interlocutors' – and his and our – igno-
rance about these fundamental things. He introduces, in
short, a new kind of conversation, about the things that
matter most to us; he tells us what he thinks those things
might be, then shows us how interestingly difficult it is
for us to come to reliable conclusions about them. So in
one sense he has no doctrine or set of beliefs that we can
adopt; and in so far as he does – the leading of a rational
and good life – he leaves us with an open, uncircum-
scribed future. This, at least, is Nehamas's case, and
whether or not it is strictly true, it enables him to say
something very interesting about Socrates and his effect
on some of the philosophers who came after him. These
philosophers, Montaigne, Nietzsche and Foucault, he
writes,

> care more about the fact that Socrates made something
> new out of himself, that he constituted himself as an
> unprecedented type of person, than about the particular
> type of person that he became. What they took from
> him was not the specific mode of life, the particular self,
> he fashioned for himself but his self-fashioning in
> general. Socrates is the prototypical artist of living

because, by leaving the process he followed absolutely indeterminate, he also presents his final product as nonbinding; a different procedure, with different materials, can create another life and still be part of his project. To imitate Socrates is therefore to create oneself, as Socrates did. But it is also to make oneself different from anyone else so far, and since that includes Socrates himself, it is to make oneself different from Socrates as well. That is why he can function as the model for the individualist, aestheticist artists of living whose main purpose is to be like no-one else.

Socrates is exemplary as a self-fashioner; he is an exemplary version of being exemplary. He doesn't say 'be like me', partly because, in Nehamas's version, there is virtually no him to be like. You just witness in Plato's texts someone doing something distinctively, and the effect he has on his interlocutors and readers. There may be Platonists but there cannot be Socratesians. 'What Montaigne learns from Socrates', Nehamas writes, 'is that to follow him is to be different from him.' That is, to be neither like him nor unlike him – as in a positive or negative conversion experience – but to be someone else, himself. And Socrates doesn't tell you how to do this; he shows you, he exemplifies it in the doing. But what he, or rather Plato, shows us is how he was himself, his manner of being himself, the way he performed who he was; Plato does this by re-presenting Socrates in a written text. And

Nehamas wants us to notice that what was also liberating for these particular philosophers is that Socrates gives us no real clue about who he had identified with in becoming himself. He hides, or conceals, or subsumes, or digests his influences – who and what formed him – so he appears to be unprecedented. Socrates is not a hero-worshipper; he doesn't boast or make claims about who or what made him. He is not always aspiring to be different, to be some-one else. This, we can say, is an alternative to conversion; or possibly a radically different form of conversion.

So, for example, when Socrates speaks of education in *The Republic*, he uses the language of what we have come to think of as conversion, but with a quite different nuance:

> Education isn't what some people declare it to be, namely, putting knowledge into souls that lack it, like putting sight into blind eyes . . . The instrument with which each person learns is like an eye that cannot be turned round from darkness to light without turning the whole body . . . Education is the craft concerned with doing this very thing, this turning around . . . Education takes for granted that sight is there, but that it isn't turned the right way or looking where it ought to look, and it tries to redirect it appropriately.

Conversion means, in the language of the *OED*, 'rotation, turning, returning'; the picture is often of reorientated perception. Socrates is saying that education is the craft that makes possible this turning around. No one is lacking anything, they are just pointing, just looking, in the wrong direction. Clearly if there is a right direction to be looking in, there must be something akin to a conversion on offer. And yet it is a peculiar species of conversion, a conversion to looking properly, but not a conversion to what you should be seeing. To be converted to the right kind of looking – looking in the right direction – tells you very little about what you might see. 'The conversion that Socrates is calling for in Book VII of *The Republic* is not one of faith,' the philosopher Nancy Bauer writes in an essay, *On Philosophical Authority*:

It is a turning from mindlessness to thoughtfulness, from dogmatism to self-scrutiny, from habit to deliberateness. To become invested in this kind of conversion is not to adopt a new outlook on the world once and for all but to understand oneself as standing in endless need of discovering what rationality demands in one's own case, in one's own thought and life . . . On Socrates' understanding the philosopher's special talent, her singular role in the world, is not to produce arguments that buttress one position or another on any given issue, arguments that can be packaged and sold to those interested in speaking on the side of reason . . . Rather, the philosopher is to understand rationality as something that each individual must discover in and for himself, in his own time, on his own terms.

Socrates, and Bauer in her commentary, are drawn to the language of conversion. In Bauer's account, which is notably complementary to Nehamas's, the whole notion of conversion is needed to describe a new kind of conversion experience. What Bauer proposes is conversion to a way of doing things – Socrates' rationality, his way of using rationality, of being rational – that is only of value if it is made almost entirely one's own: rationality, as she puts it, 'that each individual must discover in and for himself, in his own time, on his own terms'. Unlike Nehamas, she wants us to be a bit like Socrates, but she wants us to really be ourselves being a bit like Socrates. He has not exactly taught us what to believe, but how to believe. He has given us something that by definition doesn't coerce assent. You could disagree with Socrates' ideas about education and be a far more appealing person than if you disagreed with Rowan Williams's Christian account of the good life. Indeed, Socrates' rationality trades in disagreement, albeit civil disagreement. To identify with Socrates is quite different – has very different consequences – than identifying with Christ. If you identify with Christ, you will at last, to some extent, know where you are, and, to some extent, who you are (at your best and at your worst); if you identify with Socrates, you will have a way, but not a truth or a life. Or rather, the truth you will now have is that the best life is the one in which you become yourself, become rational in your own way; in which imitation is kept to a minimum, or disguised. This, at least, is what Nehamas and Bauer, in their different ways, want to persuade us of. We can choose to make choices, we can fashion ourselves by using our exemplars in particular ways. Imitation may be suicide, as Emerson wrote, but there are forms of identification that can be life-enhancing, that can give us more, not less, of the life that is ours.

We have to be converted to wanting to be ourselves, and to imagining our selves as not simply and solely – as more than – our set of identifications. How can we be like other people in a way that enables us to be more like ourselves? If Socrates is some kind of clue – some kind of intimation of a different conversion story: a story about being converted into someone not in search of conversion – we can perhaps learn as much, if not more, from those situations in which conversion is so necessary as to seem virtually automatic, beyond choice; an urgent necessity, a matter of life or death. Those life-saving identifications that can ruin people's lives. When we are converted, we are converted to what, until our conversion, somebody else wanted. But psychoanalysis shows us that we can identify before we know what we are identifying with; that we can be, in the language of the adults, converted and convert ourselves as children; so we can be converted without knowing what we have converted to; or, indeed, that we have converted. Indeed, one of the things that can happen in a psychoanalysis is that people can begin to wake up to, and think about, the unconscious identifications they have made themselves out of from childhood. Conversions that have happened before we have been able to think of them as conversions at all; before the word could possibly have existed for us. And it would be the psychoanalytic way to suggest that if what we call conversion begins in childhood, and often in desperation, this might add something to the sense we make of more common adult experiences of conversion. If childhood provides a blueprint, what are the adult revisions in conversion experiences?

IV

Experiment, don't signify and interpret!
                    Gilles Deleuze and Félix Guattari,
                                    A Thousand Plateaus

In one tradition of psychoanalysis, so-called sexual perversions are considered to be an individual's self-cure for traumas of early maternal care; it being assumed that the infant and young child both experienced and imagined unbearable conflict in his or her relationship with the mother, prompting adaptations that became in puberty, and sometimes before, perverse solutions. It should be said that perversions are taken to be aberrations or deviations from a norm that has been consensually agreed upon by a particular group of people; the difficulty in these classifications is best expressed by saying that, until quite recently, many psychoanalysts took homosexuality to be a perversion. A perversion, that is to say, in relation to the norm of heterosexuality, or heterosexual genital sex; the word is used to reinforce an essentialism, a prejudice. Without descriptions of normative sexuality there could be no meaningful use of the word perversion (which literally means 'turning aside from right or truth'). One way of describing a conversion experience, of course, would be as a change of norm, a change of basic assumptions, a radical revision of preconceptions; a turning towards right or truth. So it is worth wondering what it takes for one normative story about sexuality to be replaced by another, both culturally and personally.

But in the tradition of psychoanalysis I want to discuss, so-called perversions become like devices the child has recourse to in order to survive what is felt to be a life-threatening dependence on the mother. In this story – which runs the very real risk of being merely another mother-blaming story – the mother is the kind of person whom it is dangerous to be dependent on. In the non-mother-blaming version the mother is like this for very good reasons to do with the difficulties in her own history (mother-blaming always denies the mother a history of her own, or blames her for her history). The child cannot afford to be as passively dependent as he in fact is on this unpredictable mother; and so he constructs what in psychoanalytic language are called perverse solutions. These defences are an attempt to dehumanize the mother, with a view to controlling her in fantasy. This dehumanization, Arnold Cooper writes in *The Unconscious Core of Perversions*,

> is carried out through the use of three specific fantasies . . . These fantasies are all efforts to deny the experience of being the helpless, needy baby at the mercy of a frustrating, cruel mother. First fantasy: 'I need not be frightened because my mother is really nonexistent; that is, she is dead or mechanical, and I am in complete control.' Second fantasy: 'I need not be frightened because I am beyond being controlled by my malicious mother because I am myself nonhuman – that is, dead and unable to feel pain – or less than human, a slave who can only be acted upon rather than act.' And third: 'I triumph and am in total control because no matter what cruelty my squashing, castrating, gigantic monster mother-creature visits upon me, I can extract pleasure from it, and therefore she (it) is doing my bidding.' Different mixtures of these three unconscious fantasies . . . erase passivity by denying human maternal

control of oneself as human, by defensively converting
active to passive, and by extracting pleasure out of being
controlled. These three fantasies deny that the mother
has hurt or can hurt the child. In effect the infant says,
'(1) She doesn't exist, (2) I don't exist, (3) I force her –
now a nonhuman "it" – to give me pleasure.'

There are three conversions here, and they all occur
in fantasy, in the child's mind, because they can't exist in
any other form; the child is helpless and so can't change
reality, and has to live in his mind. The alive mother is
converted into a dead (i.e. harmless) mother, the suffering self is converted to a deadened self beyond suffering,
and the suffering the mother inflicts on that self is converted into pleasure. But the primary conversion, which
makes all the others possible, is the one Cooper describes
as 'the child defensively converting active to passive';
converting – that is to say, redescribing to himself –
what he has passively endured as something he has
actively engineered. So he is triumphantly in control. He
becomes, in fantasy, like the mother; the one who is in
charge. This is conversion as magic; when the experienced world is unbearable, it is converted, wishfully, as
it were, not simply to a bearable world but to an extremely
pleasurable one. Conversion here doesn't merely replace
one reality with another; it redeems an existing reality:
conversion as psychic alchemy. Conversion turns up
when something deemed to be unmodifiable and intoler-

able has to be transfigured. One part of the self, in this fiction, converts another part (in what could be described as a pre-figuration of later conversion scenes): the converter says, 'when you feel at your most terrified and destitute, pretend to be immensely powerful'. And the converted part of the self is gleefully megalomaniac and apparently invulnerable in its new-found empire. The convert is always somewhere a triumphant hedonist. He is getting the best pleasure.

Conversion is prompted by something unbearable, an unduly frustrating mother in this case; 'conversion' is the word because something extreme, something radical, something total, is required. A new pleasurable world has to be imaginatively created out of the ruins of an intolerable persecution. But this apparently self-invented conversion experience – or rather this childhood experience that adult theorists find themselves describing as conversion – is essentially an act of revenge. And so it is always worth wondering, as it were, where is the revenge in the conversion? What is being avenged? The psychoanalyst Robert Stoller, defining perversion as 'the erotic form of hatred', emphasizes what he takes to be the hostility in sexually perverse acts; and that we might also think of as the hostility in conversion experiences. 'The hostility in perversion', he writes, 'takes form in a fantasy of revenge hidden in the actions that make up the perversion and serves to convert childhood trauma to adult triumph.'

Why do men compulsively dominate women? Because they felt dominated by their mothers in childhood. Indeed, it would be possible, though unnecessarily crude, to see male identity – and perhaps female identity in a different way – as converting childhood trauma into adult triumph (misogyny is always triumphalist). But in this account survival depends upon what we can call a capacity for conversion. And the irony is – if irony is not too weak a word – that in this perverse solution, in this necessary conversion, the man merely turns himself into a version of the woman he hates and fears (the dominated becomes a dominatrix). The identification with the cruel, frustrating mother becomes compulsory, apparently unavoidable. And it is then all too easy for the man to blame the woman for converting him, for bringing out the worst in him, for turning him into something he would really rather not be. It is not clear, if we are to use the language of conversion, who has converted whom. In this psychoanalytic account the child is described as undergoing a forced conversion; ideally, it is assumed, he wouldn't want to have to choose, to convert to, the perverse solution.

Historically, the idea of conversion has organized or even over-organized our pictures and descriptions of dramatic and significant change in a life. We can see, even from these brief considerations, how and why we might be suspicious of conversion experiences, and seek different pictures and descriptions of how to change.

And, indeed, why conversion itself can fascinate and horrify, as another obscure object of desire. Science, of course, is conventionally proffered as the great clarifier of obscure objects of desire. And so, perhaps, in the all too familiar story, it is empirical science – what is called, in a telling phrase, evidence-based belief – that we should put our money on, and thereby acquire, at last, beliefs supposedly purged of perceptions distorted by wishes and desires. Perhaps experimentation – from the so-called experimental method of the sciences, to the more ordinary process of trying things out – is the antidote or the alternative to conversion experiences, to those momentous changes of belief that are changes of life. Or is experimenting – or even scientific method itself – just one more thing we can be converted to? People certainly don't tend to think of themselves as experimenting with conversion experiences, even when it turns out that they are.

Coda

'Plagues', Hugh Trevor-Roper remarks in passing in *The Rise of Christian Europe*, 'can be of decisive importance in history.' Whether or not the COVID virus is a plague, or like one, we are in the gradual process of discovering both what kind of decisive change it has brought to our lives, and what kind of decisions we want to make about our lives in the light of it. It is very difficult to know what kind of Europe – indeed, what kind of world – we will be left with in the aftermath of this devastation. But we can be sure that, despite people's wish and capacity to return to so-called normal life, the shock waves, the after-effects of this virus, will be felt for generations to come. Those of us who lived through it will become, among other things, the ones who lived through it.

There has been the managing of day-to-day survival, and then there is, and has been, the wholly unpredictable and deferred effect of the trauma. We have had to suspend realistic ideas about the future, but the one thing we can be certain of is a future in which we will be discovering and taking the consequences of the virus. It has been a salutary and horrifying example, if we needed one, of how much significant change in our lives is imposed upon us by both the natural world and political ideologies. It has exposed our natural vulnerability as natural creatures, and the brutal chaos of capitalist culture. It has over-exposed us to the acknowledgement of what we may be unable to change – the virus as some kind of Dar-winian super-hero, or universal acid, that we struggle to defeat and contain – and what we now feel we urgently need to change. The enormity of the threat has confronted us with, and reminded us of, what might matter to us now. The virus is both a thing in itself and an analogy for the other deadly contagions – most obviously misogyny and racism, and profiteering, our terror of sympathy and our phobia of kindness and fellow feeling – that we are daily intimidated by and don't always know how to contend with. And don't always want to.

This book was written before the devastation of the COVID virus, and before it occurred to any of us, at least in the affluent West, that such things could happen to us. It may be that when catastrophic change is inflicted upon us, with all the suffering that entails, we may become more able and willing to consider and discuss and create the kind of change we would like, the kind of change that we realize we need in order to get the lives we want. But to do that we must resist the temptation to get back to normal, now that we can see more clearly what normality has involved us in. And indeed who decides what we take normal to be.

Acknowledgements

The chapters in this book were first given, in different versions, as lectures at the University of York, and I am grateful, as ever, for the engaged responses from colleagues and students that helped me go on thinking about what I was doing. Hugh Haughton has, as always, said many illuminating things about what I'm doing. The first chapter was given as a talk at a Confer conference in London, and again to a very attentive audience.

Judith Clark has been the inspiring and reliably encouraging person who makes the writing, among many other things, possible and pleasurable.